The Colorado Mathematical Olympiad
and Further Explorations

Alexander Soifer

The Colorado Mathematical Olympiad and Further Explorations

From the Mountains of Colorado to the Peaks of Mathematics

With over 185 Illustrations

Forewords by

Philip L. Engel
Paul Erdős
Martin Gardner
Branko Grünbaum
Peter D. Johnson, Jr.
and Cecil Rousseau

 Springer

Alexander Soifer
College of Letters, Arts and Sciences
University of Colorado at Colorado Springs
1420 Austin Bluffs Parkway
Colorado Springs, CO 80918
USA
asoifer@uccs.edu

ISBN: 978-0-387-75471-0   e-ISBN: 978-0-387-75472-7
DOI: 10.1007/978-0-387-75472-7
Springer New York Dordrecht Heidelberg London

Library of Congress Control Number: 2011925380

Printed on acid-free paper

Springer is part of Springer Science+Business Media (www.springer.com)

To all those people
  throughout the world,
    who create Olympiads
for new generations of mathematicians.

Forewords to The Colorado Mathematical Olympiad and Further Explorations

We live in an age of extreme specialization – in mathematics as well as in all other sciences, in engineering, in medicine. Hence, to say that probably 90% of mathematicians cannot understand 90% of mathematics currently published is, most likely, too optimistic. In contrast, even a pessimist would have to agree that at least 90% of the material in this book is readily accessible to, and understandable by 90% of students in middle and high schools. However, this does not mean that the topics are trivial – they are elementary in the sense that they do not require knowledge of lots of previously studied material, but are sophisticated in requiring attention, concentration, and thinking that is not fettered by preconceptions. The organization in groups of five problems for each of the "Olympiads", for which the participants were allowed four hours, hints at the difficulty of finding complete solutions. I am convinced that most professional mathematicians would be hard pressed to solve a set of five problems in two hours, or even four.

There are many collections of problems, for "Olympiads" of various levels, as well as problems in a variety of journals.

What sets this book apart from the "competition" are several aspects that deserve to be noted.

- The serenity and enthusiasm with which the problems, and their solutions, are presented;
- The absence of prerequisites for understanding the problems and their solutions;
- The mixture of geometric and combinatorial ideas that are required in almost all cases.

The detailed exposition of the trials and tribulations endured by the author, as well as the support he received, shed light on the variety of influences which the administration of a university exerts on the faculty. As some of the negative actions are very probably a consequence of mathophobia, the spirit of this book may cure at least a few present or future deciders from that affliction.

Many mathematicians are certainly able to come up with an interesting elementary problem. But Soifer may be unique in his persistence, over the decades, of inventing worthwhile problems, and providing amusing historical and other comments, all accessible to the intended pre-college students.

It is my fervent hope that this book will find the wide readership it deserves, and that its readers will feel motivated to look for enjoyment in mathematics.

Branko Grünbaum
Departmernt of Mathematics
University of Washington, Seattle

Here is another wonderful book from Alexander Soifer. This one is a more-than-doubling of an earlier book on the first 10 years of the Colorado Mathematical Olympiad, which was founded and nourished to robust young adulthood by– Alexander Soifer.

Like *The Mathematical Coloring Book*, this book is not so much mathematical literature as it is literature built around mathematics, if you will permit the distinction. Yes, there is plenty of mathematics here, and of the most delicious kind. In case you were unaware of, or had forgotten (as I had), the level of skill, nay, art, necessary to pose good olympiad- or Putnam exam-style problems, or the effect that such a problem can have on a young mind, and even on the thoughts of a jaded sophisticate, then what you have been missing can be found here in plenty – at least a year's supply of great intellectual gustation. If you are a mathematics educator looking for activities for a math. club – your search is over! And with the *Further Explorations* sections, anyone so inclined could spend a lifetime on the mathematics sprouting from this volume.

But since there will be no shortage of praise for the mathematical and pedagogical contributions of *From the Mountains of Colorado...*, let me leave that aspect of the work and supply a few words about the historical account that surrounds and binds the mathematical trove, and makes a story of it all. The Historical Notes read like a war diary, or an explorer's letters home: there is a pleasant, mundane rhythm of reportage – who and how many showed up from where, who the sponsors were, which luminaries visited, who won, what their prizes were – punctuated by turbulence, events ranging from A. Soifer's scolding of a local newspaper for printing only the names of the top winners, to the difficulties arising from the weather and the shootings at Columbine High School in 1999 (both matters of life and death in Colorado), to the inexplicable attempts of university administrators to impede, restructure, banish, or destroy the Colorado Mathematical Olympiad, in 1985, 1986, 2001 and 2003. It is fascinating stuff. The very few who have the entrepeneurial spirit to attempt the creation of anything like an Olympiad will be forewarned and inspired.

The rest of us will be pleasurably horrified and amazed, our sympathies stimulated and our support aroused for the brave ones who bring new life to the communication of mathematics.

Peter D. Johnson. Jr.
Department of Mathematics and Statistics
Auburn University

In the common understanding of things, mathematics is dispassionate. This unfortunate notion is reinforced by modern mathematical prose, which gets good marks for logic and poor ones for engagement. But the mystery and excitement of mathematical discovery cannot be denied. These qualities overflow all preset boundaries.

On July 10, 1796, Gauss wrote in his diary

$$\text{EYPHKA! } num = \Delta + \Delta + \Delta$$

He had discovered a proof that every positive integer is the sum of three triangular numbers $\{0, 1, 3, 6, 10, \ldots n(n+1)/2, \ldots\}$. This result was something special. It was right to celebrate the moment with an exclamation of Eureka!

In 1926, an intriguing conjecture was making the rounds of European universities.

> If the set of positive integers is partitioned into two classes, then at least one of the classes contains an n-term arithmetic progression, no matter how large n is taken to be.

The conjecture had been formulated by the Dutch mathematician P. J. H. Baudet, who told it to his friend and mentor Frederik Schuh.

B. L. van der Waerden learned the problem in Schuh's seminar at the University of Amsterdam. While in Hamburg, van der Waerden told the conjecture to Emil Artin and Otto Schreier as the three had lunch. After lunch, they adjourned to Artin's office at the University of Hamburg to try to find a proof. They were successful, and the result, now known as van der Waerden's Theorem, is one of the *Three Pearls of Number Theory* in Khinchine's book by that name. The story does not end there. In 1971, van der Waerden published a re-markable paper entitled *How the Proof of Baudet's Conjecture was Found.*[1] In it, he describes how the three mathematicians searched for a proof by drawing diagrams on the blackboard to represent the classes, and how each mathematician had *Einfälle* (sudden ideas) that were crucial to the proof. In this account, the reader is a fourth person in Artin's office, observing with each Einfall the rising anticipation that the proof is going to work. Even though unspoken, each of the three must have had a "Eureka moment" when success was assured.

From the Mountains of Colorado to the Peaks of Mathematics presents the 20-year history of the Colorado Mathematical Olympiad. It is symbolic that this Olympiad is held in Colorado. Colorado is known for its beauty and spaciousness. In the book there is plenty of space for mathematics. There are wonderful problems with ingenious solutions, taken from geometry, combinatorics, number theory, and other areas. But there is much more. There is space to meet the participants, hear their candid comments, learn of their talents, mathematical and otherwise, and in some cases to follow their paths as professionals. There is space for poetry and references to the arts. There is space for a full story of the competition – its

---

[1] Studies in Pure Mathematics (Presented to Richard Rado), L. Mirsky, ed., Academic Press, London, 1971, pp 251-260.

dreams and rewards, hard work and conflict. There is space for the author to comment on matters of general concern. One such comment expresses regret at the limitations of currently accepted mathematical prose.

> In my historical-mathematical research for *The Mathematical Coloring Book*, I read a good number of nineteenth-century Victorian mathematical papers. Clearly, the precision and rigor of mathematical prose has improved since then, but something charming was lost – perhaps, we lost the "taste of time" in our demand for an "objective," impersonal writing, enforced by journal editors and many publishers. I decided to give a historical taste to my Olympians, and show them that behind Victorian clothing we can find the pumping heart of the Olympiad spirit. [p. 235]

Like Gauss, Alexander Soifer would not hesitate to inject *Eureka!* at the right moment. Like van der Waerden, he can transform a dispassionate exercise in logic into a compelling account of sudden insights and ultimate triumph.

Cecil Rousseau
Professor Emeritus
University of Memphis
Chair, USA Mathematical Olympiad
Committee

Forewords to Colorado Mathematical Olympiad

Love! Passion! Intrigue! Suspense! Who would believe that the history of a mathematics competition could accurately be described by words that more typically appear on the back of a popular novel? After all, mathematics is dull; history is dull; school is dull. Isn't that the conventional wisdom?

In describing the history of the Colorado Mathematical Olympiad, Alexander Soifer records the comments of a mathematics teacher who anonymously supported the Olympiad in each of its first ten years. When asked why, this unselfish teacher responded "I love my profession. This is my way to give something back to it." Alexander also loves his profession. He is passionate about his profession. And he works hard to give something back.

The Colorado Mathematical Olympiad is just one way Alexander demonstrates his love for mathematics, his love for teaching, his love for passing on the incredible joy of discovery. And as you read the history of the Olympiad, you cannot help but be taken up yourself with his passion.

But where there is passion, there is frequently intrigue. Here it involves the efforts of school administrators and others to help – or to hinder – the success of the Olympiad. But Alexander acknowledges that we all must have many friends to help us on the journey to success. And the Olympiad has had many friends, as Alexander so carefully and thankfully records.

One of the great results of the Olympiad is the demonstration that real mathematics can be exciting and suspenseful. But the Olympiad also demonstrates the essence of mathematical research, or what mathematicians really do as they move from problem to example to generalization to deeper results to new problem. And in doing so it provides an invaluable lesson to the hundreds of students who participate each year.

It is appropriate on an anniversary to look back and take stock. It is also appropriate to look forward. This book does both, for the Colorado Mathematical Olympiad is alive and well, thanks to its many ardent supporters. And for that we can all rejoice.

PHILIP L. ENGEL
President, CNA Insurance Companies
Chairman of the Board,
MATHCOUNTS Foundation
March 28, 1994, Chicago, Illinois

The author started the Colorado Olympiad in 1984, 10 years ago, and it was a complete success and it is continuing. Several of the winners have already got their Ph.D.'s in Mathematics and Computer Science.

The problems are discussed with their solutions in great detail. A delightful feature of the book is that in the second part more related problems are discussed. Some of them are still unsolved; e.g., the problem of the chromatic number of the plane – two points of the plane are joined if their distance is 1 – what is the chromatic number of this graph? It is known that it is between 4 and 7. I would guess that it is greater than 4 but I have no further guess. Just today (March 8, 1994) Moshe Rosenfeld asked me – join two points of the plane if their distance is an odd integer – is the chromatic number of this graph finite? He proved that if four points are given, the distances cannot all be odd integers.

The author states an unsolved problem of his and offers a prize of $100 for it. For a convex figure F in the plane, $S(F)$ denotes the minimal positive integer n, such that among any n points inside or on the boundary of F there are three points that form a triangle of area $\frac{1}{4}|F|$ or less, where $|F|$ is the area of F. Since for any convex figure F, $S(F) = 5$ or 6, it is natural to ask for a classification of all convex figures F, such that $S(F) = 6$.

I warmly recommend this book to all who are interested in difficult elementary problems.

Paul Erdős
Member of the Hungarian
Academy of Sciences
Honorary Member of the
National Academy of
Sciences of the USA
March 8, 1994, Boca Raton, Florida

Alexander Soifer, who founded and still runs the famous Colorado Mathematics Olympiad, is one of the world's top creators of significant problems and conjectures. His latest book covers the Olympiad's first ten years, followed by additional questions that flow from Olympiad problems.

The book is a gold mine of brilliant reasoning with special emphasis on the power and beauty of coloring proofs. Strongly recommended to both serious and recreational mathematicians on all levels of expertise.

Martin Gardner
March 10, 1994, Hendersonville,
North Carolina

Many of us wish we could contribute to making mathematics more attractive and interesting to young people. But few among professional mathematicians find the time and energy to actually do much in this direction. Even fewer are enterprising enough to start a completely new project and continue carrying it out for many years, making it succeed against overwhelming odds. This book is an account of such a rare endeavor. It details one person's single-minded and unwavering effort to organize a mathematics contest meant for and accessible to high school students. Professor Soifer managed to secure the help of many individuals and organizations; surprisingly, he also had to overcome serious difficulties which should not have been expected and which should not have arisen.

The book is interesting in many ways. It presents the history of the struggle to organize the yearly "Colorado Mathematical Olympiad"; this should help others who are thinking of organizing similar projects. It details many attractive mathematical questions, of varying degrees of difficulty, together with the background for many of them and with well explained solutions, in a manner that students as well as those who try to coach them will find helpful. Finally, the *"Further*

Explorations" make it clear to the reader that each of these questions – like all of mathematics – can be used as a stepping stone to other investigations and insights.

I finished reading the book in one sitting – I just could not put it down. Professor Soifer has indebted us all by first making the effort to organize the Colorado Mathematical Olympiads, and then making the additional effort to tell us about it in such an engaging and useful way.

Branko Grünbaum
University of Washington
March 18, 1994, Seattle, Washington

If one wants to learn about the problems given at the 1981 International Mathematical Olympiad, or to find a statistical summary of the results of that competition, the required information is contained in Murray Klamkin's book *International Mathematical Olympiads 1979–1985*. To find out about the members of the 1981 USA team (Benjamin Fisher, David Yuen, Gregg Patruno, Noam Elkies, Jeremy Primer, Richard Stong, James Roche and Brian Hunt) and what they have accomplished in the intervening years, one can read the booklet *Who's Who of U.S.A. Mathematical Olympiad Participants 1972–1986* by Nura Turner. For a view from behind the scenes at the 1981 IMO, there is the interesting article by Al Willcox, "Inside the IMO," in the September-October, 1981 issue of *Focus*. News accounts of the IMO can be found in the *Time* and *Newsweek* as well as major newspapers. However, even if one is willing to seek out these various sources, it is hard to get a full picture of such a Mathematical Olympiad, for it is much more than a collection of problems and a statistical summary of results. Its full story must be told in terms of dreams, conflicts, frustration, celebration and joy.

Now thanks to Alexander Soifer, there is a book about the Colorado Mathematical Olympiad that gives more than just the problems, their solutions, and statistical information about the results of the competition. It tells the story of this competition in direct, human terms. Beginning with Soifer's own experience as a student in Moscow, *Colorado Mathematical Olympiad* describes the genesis of the mathematical competition he has created and gives a picture of the work required to gain support for such a project. It mentions participants by name and tells of some of their accomplishments. It acknowledges those who have contributed problems and it reveals interesting connections between the contest problems and mathematical research. Of course, it has a collection of mathematical problems and solutions, very beautiful ones. Some of the problems are from the mathematical folklore, while others are striking original contributions of Soifer and some of his colleagues. Here's one of my favorites, a problem contributed by Paul Zeitz.

> Twenty-three people of positive integral weight decide to play football. They select one person as referee and then split up into two 11-person teams of equal *total* weight. It turns out that no matter who is chosen referee this can always be done. Prove that all 23 people have the same weight.

The problems alone would make this book rewarding to read. But *Colorado Mathematical Olympiad* has more than attractive mathematical problems. It has a compelling story involving the lives of those who have been part of this competition.

Cecil Rousseau
Memphis State University
Coach, U.S.A. team for the International
Mathematical Olympiad
April 1, 1994, Memphis, Tennessee

Contents

The Sixth Colorado Mathematical Olympiad

Seventh Colorado Mathematical Olympiad

Eighth Colorado Mathematical Olympiad

Ninth Colorado Mathematical Olympiad

Tenth Colorado Mathematical Olympiad

Part II. Further Explorations

Twelfth Colorado Mathematical Olympiad

Thirteenth Colorado Mathematical Olympiad

Fourteenth Colorado Mathematical Olympiad

Fifteenth Colorado Mathematical Olympiad

Sixteenth Colorado Mathematical Olympiad

Seventeenth Colorado Mathematical Olympiad

Eighteenth Colorado Mathematical Olympiad

Nineteenth Colorado Mathematical Olympiad

Twentieth Colorado Mathematical Olympiad

Part IV. Further Explorations of The Second Decade

Greetings to the Reader: Preface to The Colorado Mathematical Olympiad and Further Explorations

Beauty is an instance which plainly shows that culture is not simply utilitarian in its aims, for the lack of beauty is a thing we cannot tolerate in civilization.
– Sigmund Freud, 1930, Civilization and Its Discontents[2]

Talent is not performance; arms and legs are no dance.
– Hugo von Hofmannsthal, 1922, Buch der Freunde[3]

Imagination is more important than knowledge. For knowledge is limited to all we now know and understand, while imagination embraces the entire world, and all there ever will be to know and understand.
– Albert Einstein

I have been often asked: what are Mathematical Olympiads for? Do they predict who will go far in mathematics and who will not? In fact, today, in 2010, in Russia, Olympiads are *officially* used as predictors of success, for those who placed high enough in important enough Olympiads get admitted

---

[2] [F]
[3] [Ho]

to important enough universities! Having spent a lifetime in Olympiads of all levels, from participant to organizer, I have certainly given these kinds of questions a lot of thought.

There is no way to do well by accident, by luck in an Olympiad (i.e., essay-type competition, requiring presentation of complete solutions). Therefore, those who even sometimes have done well in Olympiads, undoubtedly have talent. Does it mean they will succeed in mathematics? As the great Austrian writer Hugo von Hofmannsthal put it in 1922, *Talent is not performance; arms and legs are no dance.* With talent, it still takes work, hard work to succeed. I would say talent imposes an obligation on its owner, a duty not to waste that talent. I must add, nothing is a guarantee of success, life interferes, throws barriers on our way. It is critical to know in your gut that in order to succeed no reason for failure should be acceptable.

The inverse here is not true. Those who have not done well in Olympiads do not necessarily lack talent. They may be late bloomers – Einstein comes to mind. Their talent may be in another endeavor. I believe there are no talentless people – there are people who have not identified their talent or have not developed it.

In addition to allowing talents to shine, Olympiads introduce youngsters to a kind of mathematics they may not have seen in school. It was certainly the case with me. Olympiads showed me the existence of mathematics that justifies such adjectives as beautiful, elegant, humorous, defying intuition. Olympiad mathematics inspires and recruits; it passes the baton to a new generation of mathematicians.

And one more thing, which apparently I said long ago (see my newspaper interview in *Historical Notes* 14). Students who do well in Olympiads have freedom of thought. They can look at usual things in a new way, as great painters and poets do. Olympiads provide students with an opportunity to express and celebrate this creative freedom.

The Colorado Mathematical Olympiad has survived for well over two decades. Happy anniversary to the Olympiad and *all* those who made it possible! The Olympiad has become *a part of Colorado life* for many students, parents and schools in Colorado. I know that the Olympiad is approaching when I receive e-mails from eagerly awaiting participants, their parents and teachers.

It has been reasonable to limit participation to students of the State of Colorado. However, I could not refuse guest participation, and we have had Olympians traveling from Long Island, New York, Kansas, and even the national team of the Philippines. For a number of years, winners of the Mobile, Alabama mathematics competition have been receiving as a prize a visit to compete in the Colorado Mathematical Olympiad.

I have just now noticed that in preparing for Springer new (expanded) editions of my books *Mathematics as Problem Solving*, *How Does One Cut a Triangle?*, and *Geometric Etudes in Combinatorial Mathematics*, I preserved original books within new editions, and added new parts within new editions. I could not alter the wholeness of the first editions because as fine works of art, these original books had to be preserved unaltered. This book dramatically expands upon the 1994 original *Colorado Mathematical Olympiad: The First Ten Years and Further Explorations*. But the original is essentially preserved in the present book as Parts I and II. Of course, I have added some new mathematical results in Part II, for mathematics did not stay put for the intervening 16 years. I have also added a number of historical sketches, for after *The Mathematical Coloring Book* [S11] my interests have broadened and the writing style has changed to accommodate historical writing and storytelling. The original Parts I and II are followed by the brand new Parts III and IV and the new Part V, *Winners Speak: Reminiscences in Eight Parts*.

The original Part I, presents the history of the first ten years of the Olympiad. It also presents all 51 problems of the first ten years of the Olympiad with their solutions. History, Problems, and Solutions sections are organized by the year of the Olympiad. Problem number *i.j* indicates problem *j* of the *i*-th Olympiad. After Historical Notes, Problems, and Solutions come *Further Explorations*, a unique feature of this book, not found in any of the numerous books reporting Russian, American, Chinese, International, and other Olympiads. Each Exploration takes off from one or more Olympiad problems presented in this book, and *builds a bridge to the forefront of mathematics*, in some explorations to open problems of mathematics. This is the feature the great mathematician Paul Erdős liked the most, when he decided to write a foreword for the 1994 edition of the book.

The new Part III presents the history and problems of the second decade. I can now see how the problems of the Olympiad have matured over the years. Part IV then offers 10 new *Further Explorations*, for the total of 20 bridges between Olympiad problems and problems of real mathematics. I mean here "real" in the sense of real mathematicians working on these kinds of problems, riding these kinds of trains of thought – and not in a sense of "real" life. In 1921 Albert Einstein addressed the correlation between mathematics and reality in a most convincing way: *As far as the propositions of mathematics refer to reality, they are not certain; and as far as they are certain, they do not refer to reality.*

Parts I, II, III, and IV are followed by Part V, *Winners Speak: Reminiscences in Eight Parts.* In this part, several winners of the Olympiad evaluate the role of the Olympiad in their lives, and describe their young professional careers – life after the Olympiad.

I now wish to express my gratitude to many people who allowed me to make the Colorado Mathematical Olympiad a reality. First of all I thank the many thousands of young mathematicians, Colorado Olympians, for without them, their dedication to the Olympiad, their efforts and their demonstrated brilliance all our hard work would make no sense.

I am grateful to my Dean of Letters, Arts and Sciences Tom Christensen and my Chancellor Pam Shockley-Zalaback for establishing the *Admission Window* for the Olympiad's medalists, the window equal to that for the USA Olympic sportsmen (who happen to live and train in Colorado Springs); and to Pam for proposing and funding *Chancellor's Scholarships for Olympiad's Medalists*. I thank the two of them and my chair Tom Wynn for understanding and appreciating the intensity of my ongoing work on the Olympiad for the past 27+ years.

I thank all those who volunteered their time and talent to serve as judges and proctors of the Olympiad, especially to those judges who have served the longest: Jerry Klemm, Gary Miller, Bob Ewell, Shane Holloway and Matt Kahle. I am grateful to the Olympiad's managers who gave long months every year to organizing the Olympiad and its registration: Andreanna Romero, Kathy Griffith, and Margie Teals-Davis. I thank people of Physical Plant, Media Center, and the University Center, and above all Dave Schnabel, Christian Howells, Rob Doherty, Mark Hallahan, Mark Bell, and Jeff Davis.

The Olympiad has been made possible by dedicated sponsors. I thank them all, and first of all the longest-term sponsors Dr. Stephen Wolfram and Wolfram Research; CASIO; Texas Instruments; Colorado Springs School Districts 11, 20, 2, and 3; Rangely High School; Chancellor's and Vice Chancellors' Offices and the Bookstore of UCCS. I am infinitely indebted to Greg Hoffman, whose five companies all in turn became major contributors most definitely due to Greg's trust in and loyalty to the Olympiad.

Every year Colorado Governors Roy Romer, Bill Owens, and Bill Ritter, as well as Senator and later Director of Labor Jeffrey M. Welles have written congratulatory letters to the winners of the Olympiad. The most frequent speakers at the Award Presentation Ceremonies were Chancellors Dwayne Nuzum and Pamela Shockley-Zalaback, Deans James Null and Tom Christensen, sponsor Greg Hoffman, Senator Jeffrey M. Welles, and Deputy Superintendants Maggie Lopez of Air Academy District 12 and Mary Thurman of Colorado Springs District 11. My gratitude goes to all of them and all other speakers of the Award Presentations.

I thank the long term senior judge of the Olympiad, Dr. Col. Robert Ewell and our former graduate student Phil Emerick for translating some of my hand-drawn illustrations into sharp computer-aided designs.

I am forever indebted to all my fabulous mathematics teachers, from first grade to Ph.D. program: Matilda Koroleva, Klara Dimanshtein, Nikolaj Konstantinov, Tatiana Fideli, Ivan Morozkin, Yuri Mett, and Leonid Kulikov. The value and influence of my lifelong interactions with other Olympiad people are impossible to overestimate – in fact, this book is dedicated "to all those people throughout the world who create Olympiads for new generations of mathematicians."

My late parents, the artist Yuri Soifer and the actress Frieda Gofman, gave me life and filled it with magnificent art; the longer I live the more I realize how much this environment has influenced and formed me. They rose to the occasion, recognized my enthusiasm toward mathematics, and respected my switch from the art of music to the art of mathematics. My parents and my kids, Mark, Julia, Isabelle and Leon have been my love and inspiration. They all participated in the Olympiad and some won awards. My veteran judges and I will never forget how for years tiny beautiful Isabelle in a Victorian dress passed the Olympiad's buttons to the winners. I owe you all so much!

I admire the great talent and professionalism, taste and kind attention of the first readers of this manuscript, who are also the authors of this book's forewords: Branko Grünbaum, Peter D. Johnson, Jr., and Cecil Rousseau. Thank you so very much!

I thank my Springer editor Elizabeth Loew for constant support and good cheer, and Ann Kostant for inviting this and my other seven books to the historic Springer. I am grateful to Dr. David P. Kramer of Massachusetts for a superb copy-editing, and to Max Mönnich of Berlin for designing a wonderful cover.

The Olympiad weathered bad weather postponements and administrators impersonating barricades on the Olympic track. The Olympiad also encountered enlightened administrators and many loyal volunteers. It is alive and well. I may meet with you again soon on the pages of a future book inspired by the third decade of the Olympiad!

Alexander Soifer
Colorado Springs
December 2010

Preface to Colorado Mathematical Olympiad

> *Mathematics, rightly viewed, possesses not only truth, but supreme beauty... capable of a stern perfection such as only the greatest art can show.*
>
> — Bertrand Russell

> *It has been proved by my own experience that every problem carries within itself its own solution, a solution to be reached by the intense inner concentration of a severe devotion to truth.*
>
> — Frank Lloyd Wright

The Colorado Mathematical Olympiad has survived for a decade. Happy anniversary to the Olympiad and *all* people who made it possible!

What is the Olympiad? Where does one get its problems? How to solve them? These questions came through the years from every corner of Colorado and the world. This is the right time to reply: we have accumulated enough of striking history, intriguing problems and surprising solutions, and we have not forgotten too much yet.

Part I, *The First Ten Years,* is my reply to these questions. It describes how and why the Olympiad started, and gives a fairly detailed history of every year. Many, although inevitably far from all, people who made the Olympiad possible are recognized: sponsors, problem creators, judges, proctors, and, of course, contestants.

Some people who tried to make the Olympiad *impossible* are mentioned as well. The history of human events is never a bed of roses. Future organizers of Olympiads need to know that they will face support and understanding from some and indifference and even opposition from others.

Part I also presents all 51 problems of the first ten years of the Olympiad with their solutions. Two or even three solutions are presented when as many different and beautiful solutions have been found. Historical notes, Problems, and Solutions sections are organized by the Olympiads. Accordingly problem *i.j* stands for problem *j* of the *i*-th Olympiad.

I did enjoy reliving the ten years of the Olympiad and revisiting all its problems (nearly all these solutions were *written* for the first time for this book). Yet, in working on this book my real inspiration came when the idea for Part II: *Further Explorations* occurred to me. This is when I began to feel that I was writing a *mathematical* book.

Part II in its ten essays demonstrates what happens in a mathematical exploration when a problem at hand is solved. Each essay takes an Olympiad problem (or two, or three of them) and shows how its solution gives birth to deeper, more exciting, and more general problems. Some of these second generation problems are open (i.e., not solved by anyone!). Others are solved in this book. Several problems are left unsolved, even though I know beautiful solutions, to preserve for the reader the pleasure of discovering a solution on his own. In some essays the reader is led to the third generation problems. Several open problems carry a prize for their first solutions. For example, this is the first time that I am offering $100 for the first solution of Problem E5.8 (this number refers to the eighth problem of chapter E5 of *Further Explorations*).

To the best of my knowledge, this is *the first* Olympiad problem book with a whole special part that bridges problems of mathematical Olympiads with open problems of mathematics. Yet, it is not a total surprise that I am able to offer ten mini-models of mathematical research that originate from Olympiad problems. The famous Russian mathematician Boris N. Delone once said, as Andrei N. Kolmogorov recalls in his introduction to [GT], that in fact, "a major scientific discovery differs from a good Olympiad problem only by the fact that a solution of the Olympiad problem requires 5 hours whereas obtaining a serious scientific result requires 5,000 hours."

There is one more reason why *Further Explorations* is a key part of this book. Olympiads offer high school students an exciting addition and alternative to school mathematics. They show youngsters the beauty, elegance, and surprises of mathematics. Olympiads celebrate achievements of young mathematicians in their competition with each other. Yet, I am not a supporter of such competitions for university students. University is a time to compete with the field, not with each other!

I wish to thank here my junior high and high school mathematics teachers, most of whom are not with us any more: Klara A. Dimanstein, Tatiana N. Fideli, Ivan V. Morozkin, and Yuri F. Mett. I just wanted to be like them! I am grateful to Nikolai N. Konstantinov for his fabulous mathematics club that I attended as an eighth grader, and to the organizers of the Moscow University Mathematical Olympiad: they convinced me to become a mathematician. I am grateful to the Russian Isaak M. Yaglom and the American Martin Gardner for their books that sparked my early interest in mathematics above all other great human endeavors.

The wonderful cover of this book was designed by my lifelong friend Alexander Okun. Thank you Shurik for sharing your talent with us. I thank David Turner and Mary Kelley, photographers of *The Gazette Telegraph*, for the permission to use their photographs in this book, and *The Gazette* editors for the permission to quote the articles about the Olympiad.

I am grateful to Philip Engel, Paul Erdős, Martin Gardner, Branko Grünbaum and Cecil Rousseau for kindly agreeing to be the first readers of the manuscript and providing me with most valuable feedback. I am truly honored that these distinguished people have written forewords for this book.

I applaud my Dean James A. Null for bringing about such a climate in our College of Letters, Arts and Sciences, that one is free and encouraged to create. I am grateful to my wife Maya for sustaining an intellectual atmosphere under our roof, proof-reading this manuscript, and pushing me to a pen in my less inspired times. I thank Steven Bamberger, himself an award winner in the Fourth Colorado Mathematical Olympiad, for deciphering my manuscript and converting it into this handsome volume.

The Colorado Mathematical Olympiad has survived for a decade. Happy Anniversary to the Olympiad and *all* people who made it possible!

Alexander Soifer
Colorado Springs
January 1994

Part I
The First Ten Years

Colorado Mathematical Olympiad:
How it Started and What it has Become

The Colorado Mathematical Olympiad. Why did I create it? Who needs it? Who makes it possible? To answer these questions, I have to start with my own story and a critical role mathematical Olympiads played in it.

When I was six, my mother decided that I would be a pianist. For the next nine years I attended a music school in the class of the incomparable Boris F. Platoff, as well as completed a grade and a middle school. I was 15. It was time to apply for a music college when I announced to my parents my decision to become a mathematician, and to start by entering a mathematical school. Why did I choose mathematics? It was certainly not due to school mathematics. I equally liked (and disliked!) all disciplines.

My middle school mathematics teacher K. A. Dimanstein introduced me to Moscow City and Moscow University Mathematical Olympiads, and I immediately fell in love with those problems, full of beauty, elegance, good sense of humor, and surprises.

Then books by Isaak M. Yaglom, Martin Gardner, and others entered my life. Finally, on Saturdays I attended an unforgettable mathematics club at the "Old Building" of Moscow State University offered to a group of eighth graders by Nikolai N. Konstantinov. [Added in 2010: Konstantinov remains today one of the greatest inventors of mathematical work with talented students in the world.] The club offered fabulous Olympiad-type problems and a new atmosphere, unmatched by any school. For example, I told Konstantinov with embarrassment, that I had solved a problem he had given us, but it took me over a month. To my surprise, he enjoyed the most not that I was only the second student to solve it, and not that my solution was new, but that I did not give up in the course of the month!

A. Soifer, *The Colorado Mathematical Olympiad and Further Explorations: From the Mountains of Colorado to the Peaks of Mathematics*, DOI 10.1007/978-0-387-75472-7_1, © Alexander Soifer, 2011

My magnet mathematical high school #59 near Old Arbat in the heart of Moscow delivered me the pleasure of learning math with absolutely remarkable teachers Tatiana Ivanovna Fideli and Ivan Vasilievich Morozkin. The latter even had me teach my own ninth and tenth grade classes when he was not well for extended periods of time – Morozkin trusted me to do it rather than other math teachers. Then Yuri Fomich Mett entered my life. He was a lecturer for hundreds of senior students preparing to take entrance exams at Moscow State University. Mett noticed something special in me and chose to work with me weekly one on one, on an incredible number of problems with an amazing speed.

Mathematical Olympiads have been a part of my life ever since. As a high school student, I participated in numerous Olympiads. Then came time to co-organize Olympiads for others, from running the Olympiad for my class in the famous Mathematical School #2 to serving as a judge in the Soviet Union National Olympiads.

My exodus from Russia in 1978, among other things, interrupted my participation in Olympiads. After a short stay in Boston, I arrived in Colorado Springs eager to participate in a local Olympiad. To my disbelief, I found none. Moreover, many of my peers were not particularly eager to start an Olympiad. "I do not wish to spend my time on a service activity because service matters very little in tenure decisions," replied a lady peer [Prof. Laurel Rogers] to my invitation to organize an Olympiad together.

Those who participated in many Olympiads (and won at least some of them) know how contagious Olympiads are. You get an urge to be a part of them. For those of us who can no longer participate, a new mission of creating problems and organizing Olympiads becomes a virtual necessity. It is a great responsibility and pleasure to pass the baton to new generations of mathematicians!

I came from the mathematical culture of Moscow, where the greatest mathematical minds, such as A. N. Kolmogorov, P. S. Aleksandrov, I. M. Gel'fand, E. B. Dynkin, A. Kronrod, and many others, too numerous to mention, were flaming torches for us, future generations of mathematicians. What could I do here in the United States in the beautiful Rocky Mountains, without anything resembling the great Russian mathematical traditions? A surprising solution came to me one day in the fall of 1983.

While teaching a junior course on number theory, I shared with my twenty students an idea of starting the Colorado Springs Mathematical Olympiad, and passed around a sign-up sheet. To my great surprise, some fifteen students signed up! The Olympiad became a realistic proposition: I had a group of organizers-judges. What about participants?

Anne Thrasher, the mathematics coordinator of our largest Colorado Springs School District 11, arranged for me to meet with the mathematics departments' chairs of the five high schools. Five chairs were looking at an outsider, me, with suspicion. An hour later, however, everyone smiled. The Olympiad was on. The issue of prizes, however, the ice melted and remained.

In early January 1984 I ran into Tom Burniece, the Engineering Manager of the Digital Equipment Corporation (DEC). He was excited about the idea of the Olympiad and asked me what prizes I had in mind for the winners.

"Books, certificates," I replied.

"You need to award scholarships to the top winners," Tom suggested.

"Yes, but who will donate the funds?"

"DEC people and other good folks in the industry."

Tom kept his promise. Digital donated $1,000 for the first prize winner. Incredibly, the first prize winner was the son of a Digital engineer! (See the *First Colorado Mathematical Olympiad* for details.) My Dean of Engineering, James Tracey, agreed to raise another $1,000 and a few calculators under the condition that I and other organizers of the Olympiad would not contact industries ourselves (and thus would not interfere with his raising millions for the equipment for the new Engineering Building).

Now we needed a host for the Olympiad. I hoped that my College of Engineering would welcome participants, but my dean refused. I then met with the Dean of Education, Dennis Mithaug. I told him about the Olympiad and added:

"Dennis, you are the last person at the University with whom I am discussing the Olympiad. If you refuse to be the host, I will hold the Olympiad at the City Hall, or a conference room of Hewlett-Packard, or Doherty High School. It *will* take place."

"You convinced me. We are the poorest college, but we will supply whatever you need. We are in!" said Dennis.

How can an Olympiad be evaluated by a person who has not participated in the Olympiad or its organizing, who has not witnessed excited faces of participating kids and the joy of reading solutions of future mathematicians? Only by the Olympiad's problems. We needed the least standard, the most interesting, decisively unknown to participants, problems. We did not have a large enough organization to consider an alternative of offering the competition separately for every grade. Thus, we decided to offer the same problems to everyone who came, from junior high school students to seniors. This approach made finding acceptable problems much more difficult, but positively improved the quality of the problems. We had to create problems that required for their solutions minimum of knowledge and a great deal of creativity, ingenuity and analytical thinking.

Problem 2 for the First Olympiad was contributed by the young visiting German mathematician Jörg Stelzer. Problem 3 was borrowed from the Russian mathematical folklore. Problems 1 and 4, created by me, were based on known ideas. My favorite, problem 5, was a totally original problem created in 1971-1972 by my university classmate Semjon Slobodnik and myself. (In fact, Semjon and I went back all the way to attending Konstantinov's club together, when we both were eighth-graders!)

We wanted to ensure that a good number of participants would solve something. On the other hand, we needed to clearly separate and order the winners. Hence, we selected problems of greatly varying difficulty, and arranged them from the easiest problem 1 to the most difficult problem 5. This approach worked so well that we have kept it ever since.

Everything was ready. The First Colorado Springs Mathematical Olympiad took place on April 27, 1984.

In the years that followed the First Olympiad, I tried to bring the Olympiad problems closer to "real" mathematics: many problems required construction of examples (rather than just analytical reasoning), some comprised a sequence of problems increasing in difficulty, leading to generalizations, and deeper results. These problems were meant to demonstrate ways in which mathematical research worked. I presented these ideas in my talk *"From Problems of Mathematical*

Olympiads to Open Problems of Mathematics" at the VI International Congress on Mathematical Education in August 1992 in Quebec City. I discovered there that a number of my colleagues, such as Vladimir Burjan from Slovakia and Nikolai Konstantinov from Russia, shared this belief. Konstantinov took it to the fabulous extreme by organizing a competition in which participants chose one problem from a short list and worked on it for a week [International Tournament of Towns]!

The structure of the Olympiad has survived the first decade without any changes. Here is how it works.

Our goal is to allow every student who so desires to participate in the Olympiad. To the best of my knowledge, this is the only major essay-type competition in the United States that does not restrict participation. Even the USA Mathematics Olympiad limits the essay-type part to a mere 150 participants. We offer only *individual competition* and put no restrictions on the number of participants from a school, for doing otherwise would have put schools in a position to decide who is to represent them, which would in many cases boil down to choosing polite "A" students, and not the best talents which is what we are after. It suffices to mention that twice the Olympiad was won by a student who at the time had a "C" in geometry in his school.

Young Olympians are offered 5 problems and 4 hours of time to solve them. We give no credit for answers submitted without supporting work. Conversely, a minor error that leads to an incorrect answer does not substantially reduce the credit. *Our Olympiad is genuinely and principally an essay-type competition.* We generously reward originality and creativity. Sometimes students come up with new ideas and solutions not previously known to me or anyone else!

With participation growing through the years and reaching close to 1,000, the judges, most of whom are my students and alumni, read close to 5,000 essays. Moreover, each essay is evaluated by at least two judges independently, and all disagreements are discussed by many, if not all, of the 12 to 18 judges. This enjoyable but hard job takes us usually a weekend. At the end we learn the names of the students to whom we awarded prizes (judges evaluate essays under code without knowing the names of their authors).

A week later the Award Presentation Ceremony offers a review of solutions, a lecture on mathematics, and speeches by dignitaries and legislators. Our wonderful sponsors, who believe and invest in the future of the United States, make it possible for us to award several thousand dollars in scholarships and prizes.

In the chapters that follow we will look at the history of every year of the Olympiad, as well as its problems, solutions, and *Further Explorations*.

The participation and geography of the Olympiad grew rapidly. We never restricted participation to the city of Colorado Springs, and in 1989 we formally changed the name of this state-wide event from Colorado Springs to *Colorado Mathematical Olympiad*.

I would like to mention here that there are many types of mathematical competitions throughout the world. Some expect participants to merely state answers, others are multiple choice competitions. Some competitions are oral, and completed in a matter of a couple of hours. Others go on for a week or weeks. Over the past 100+ years throughout the world the word *"Olympiad"* came to mean the particular type of competition where complete essays are expected for every problem and adequate time is offered to participants. The Colorado Mathematical Olympiad has become the largest in-person essay-type mathematical competition in the United States, a true effort of all sectors of our community.

From the bottom of my heart I thank *all* who contributed to the Olympiad through its first ten years. It is impossible to mention everyone. Nevertheless, I would like to take this opportunity to thank a good number of major contributors:

Hosts: University of Colorado Deans Dennis E. Mithaug, James A. Null, Margaret A. Bacon, and James H. Tracey.

School District Sponsors: Colorado Springs (#11), Air Academy (#20), Harrison (#2) and Widefield (#3).

Industrial and Government Sponsors: Hewlett-Packard, Digital Equipment Corporation, Texas Instruments, Casio, UCCS Bookstore, Robert Penkhus Motors, Colorado Springs City Council, United States Air Force Space Command and NORAD, and United States Space Foundation.

Higher Education Sponsors: Chancellors Neal Lane, Dwayne C. Nuzum and the University of Colorado at Colorado Springs; Deans David Finley, Timothy Fuller and the Colorado College.

Individual Sponsors: John Wicklund, John and Ardyce Putnam, Dr. David Finkleman, and Dr. William H. Escovitz.

School Sponsors: St. Mary's, Colorado Springs, Gateway (Aurora), Cheyenne Mountain, Woodland Park, Arapahoe (Littleton), West Grand (Kremling), Fruita Monument (Grand Junction).

Hotels that provided complimentary accommodations for contestants from far corners of Colorado: Antlers, Red Lion Inn, Hilton, Quality Inn.

Award Presentation Speakers: Colorado State Senators Jeffrey M. Wells and Michael Bird; Mayor of Colorado Springs Robert Isaac; Members of the Colorado Springs City Council Frank Parisi, Mary-Ellen McNally, Lisa Are and John Hazlehurst; University of Colorado Chancellors Neal Lane and Dwayne C. Nuzum; Deans Dennis E. Mithaug, Margaret Bacon, James A. Null and James H. Tracey; Colorado College Dean David Finley; Digital Equipment Corporation managers Robert Rennick and Karen Pherson; Past President of the Colorado Mathematics Teachers Association John Putnam; Texas Instruments manager Gay Riley–Pfund; President of Penkhus Motor Company Robert Penkhus; Joy Muhs of Casio.

Problem Creators: Paul Erdős (Hungarian Academy of Sciences), Joel Spencer (Courant Institute of Mathematics), Paul Zeitz (University of California, Berkeley), Steven Janke (Colorado College), Jörg Stelzer (Germany), Vladimir Baranovski (Russia), Boris Dubrov (Belarus), Alexander Trunov (Dallas), and Alexander Soifer (University of Colorado).

Judges: Gary Miller, Jerry Klemm, Lossie Ortiz, Dr. Robert Ewell, Donald Sturgill, Paul Zeitz, Keith Mann, Dr. James Modeer, Dr. Steven Janke, Dr. John Watkins, Dr. Gene Abrams, George Berry, Zoa Weber, Pam Shannon, Wanda Madrid, Dr. Jerry Diaz, Helen Muterspaugh, Stephanie Kozak, Richard Jessop, Tom Bushman, Shirley King, Regina Aragón, Mike Lechner, Michael Carpenter, Brad Johnson, Pamalee Brown, Graham Peirce, Dr. Bill Escovitz, Mark Robins, Bob Wood, Charles Phipps, Connie Parrish, Donna Yowell, Bill Young, Andrai Hak, Matt Kahle, Rhet Rodewald, Bill Walker, Tracey Tilman, Donald Holliday, Jeff Sievers, Dona Tannehill, Tom Pinson, Dr. Gregory Morrow, Dr. Dennis Eagan, Bill Walker,

Darla Waller, Maya Soifer, Edward Pegg Jr., Deanna Clement, Kelly Thompson, Dr. Graham Hawtin, Michelle Anderson, Jesse Watson, George Bajszár Jr., D'Ann DiMuccio, Ulf Gennser, and Alexander Soifer.

Administrators, Proctors: Robert Hall, Frank and Roberta Wilson and the Palmer High School Alumni Association, Anne Thrasher, Angela Schmidt, Charlotte Brummer, Bill Scott, Sharon Sherman, D'Ann DiMuccio, Steven Bamberger, John Putnam, John Wicklund, Vic Kuklin, Judy Williamson, Gloria Boken, Theodora Gonzales, Gary Riter, Laura Kadlecek, Linda Long, John Muller, Warner Nelson, Donna Binkly, Carolyn Fisher, Marjorie Card, Liz Holstein, Helen Holmgren, George Daniels, Kris Hubersburger, Jody Frost, Laura Brandt, Michael Sansing, Gwen Rosentrater, Donna Roberts, Linda Taylor, Rosemary Hetzler, Richard Rathke, Alan Versaw.

UCCS Personnel: Mark Hallahan and the Media Center, Doris Skaggs and the Physical Plant, Lynn Scott, Andrea Williams, Georgia Macopulous, Kris Fisher, Randy Kouba, Leslie O'Hara, Gayla Gallegos and the University Center.

Media: the many journalists, photographers, and editors of *The Gazette Telegraph*, who provided much needed remarkable coverage of the Olympiad.

I wish to thank teachers who produced especially many winners every year: Judy Williamson, Gary Riter, Vic Kuklin, John Wicklund, John Putnam, Lossie Ortiz, and Laura Kadlecek.

Three Celebrated Ideas

In the Colorado Mathematical Olympiad the same problems are offered to every participant from a seventh grader to a senior. This is why they must require minimal bits of knowledge for their solutions, such as the sum of angles in a triangle is equal to 180°, or the three bisectors of a triangle have a point in common. These problems do require a great deal of common sense, creativity, and imagination. Some of the problems model mathematical research: they would capitulate only to experimenting with particular cases, followed by noticing a pattern, followed in turn by generalization, formulation of a hypothesis, and finally by a proof.

Knowledge of the following three celebrated ideas of proof would be a definite plus for the reader.

Arguing by Contradiction

To argue by contradiction means to assume the conclusion is not true and from there draw a chain of deductions until you arrive at something that contradicts what is given or what is known to be true.

Pigeonhole Principle (Also Known as Dirichlet's Principle)

Let k and n be positive integers. If $kn + 1$ pigeons sit in n pigeonholes, then at least one of the holes contains at least $k + 1$ pigeons in it.

A. Soifer, *The Colorado Mathematical Olympiad and Further Explorations: From the Mountains of Colorado to the Peaks of Mathematics*, DOI 10.1007/978-0-387-75472-7_2, © Alexander Soifer, 2011

Principle of Mathematical Induction

If the first person in line is a mathematician, and every mathematician in line is followed by a mathematician, then everyone in line is a mathematician.

More seriously: Given a subset N_1 of the set of all positive integers N. If both of the following conditions are satisfied:

(1) The number 1 belongs to N_1.
(2) If the number n belongs to N_1, then $n + 1$ belongs to N_1 as well.

Then the subset N_1, in fact, coincides with the whole set N.

You will find more information about these ideas, as well as a number of problems, both solved and unsolved, in my book *Mathematics as Problem Solving* ([S1], pp. 7–23, or its new expanded Springer edition [S16]) and article [SL].

First Colorado Mathematical Olympiad
April 27, 1984

Historical Notes 1

On April 27, 1984, the few (170 to be exact), the curious, and the brave showed up at the University of Colorado at Colorado Springs (UCCS). They did not know what to expect. Neither did the journalists from the newspapers and television stations. Here is how Julie Bird described the event the next morning on the front page of the Metro Section of *The Gazette Telegraph*.

> The five problems that 170 junior high and high school students tried to solve Friday were a little bit tricky, acknowledged Olympiad organizer Alexander Soifer, a mathematics professor at the University of Colorado at Colorado Springs [UCCS].
>
> But some of the students thought the problems were a little more than tricky.
>
> As Rampart High School sophomore Tracy Lintner said, "it was hard!"
>
> Soifer said the story problems were intended to test the creative abilities of the young math minds. All work had to be shown, according to the rules, but a minor error leading to an incorrect answer would not substantially reduce their credit.
>
> "So here is the test," Soifer said after explaining the answers to other faculty and UCCS students who were to judge the students' answers. "We have four easy but by no means trivial problems, and one difficult by any means problem."

A. Soifer, *The Colorado Mathematical Olympiad and Further Explorations:*
From the Mountains of Colorado to the Peaks of Mathematics,
DOI 10.1007/978-0-387-75472-7_3, © Alexander Soifer, 2011

The question was deceivingly short:

"Forty one rooks are placed on a 10×10 checkerboard. Prove that you can choose five of them which are not attacking each other. (We say that one rook 'attacks' another if they are in the same row or column of the checkerboard.)"

Soifer's judges, all advanced mathematicians, needed a second explanation of the answer to that one. Even the second explanation didn't help a *Gazette Telegraph* reporter much.

"I wanted to give problems that some faculty could not solve, but that some grammar school wonder kids could," Soifer said.

He apparently succeeded.

Kim Roberts, a 14-year-old freshman at Rampart High School, said she "didn't really understand the question" on No. 2, which had to do with connecting dots by single wires that would allow electricity to flow only one way.

But after reviewing the first question for a while, she said she "tried a whole bunch of ways before I realized it was easy."

In that question, a tailor who enjoyed cutting cloth started out with 10 pieces of cloth. He cut one of those pieces into 10 pieces, then cut some of those pieces into 10 pieces, and continued until he thought he had 1984 pieces of cloth.

The competitors were to prove that the tailor's count was incorrect.

"Is it difficult?" he asked his judges. "I think not. Is it difficult to solve? I hope so. You see, the problem allows solution on one line."

Gaard said none of the problems was easy, although each required "logical reasoning."

"On a few I had an idea what to do, but on the others I didn't know which way to go," he said. "I'm going to show these to my uncle – he's got a master's in math – and see what he can do with them."

Soifer's judges will review the papers during the next week. The students are invited to return Friday when the solutions are reviewed and the winners are announced.

This was probably the first time that challenging mathematical problems appeared in *The Gazette Telegraph*. I received solutions of the rook problem from readers of all walks of life. Two remarkable solutions came from Digital software engineers. Jein-Shen Tsai discovered the solution that the problem authors (Slobodnik and myself) had in mind (see *First Solution* of Problem 1.5). George W. Berry found a totally new, long, but absolutely gorgeous solution (See *Third Solution* of Problem 1.5). I was so impressed that I immediately invited both of them to join the judges of the Olympiad the next year.

Right to left: Martin Puryear (second prize), Russel Shaffer (first prize), and Chris Bounds at the First Olympiad. *Photograph by David Turner, Gazette Telegraph*

Meanwhile we read, with pleasure, the essays of the competitors. One paper impressed us the most. Not only did Russel Shaffer, a senior at Doherty High School in the class of Charlotte Brummer, match everyone else by writing clear solutions of Problems 1 through 3, he was the only one to solve Problem 5. Moreover, his short, fabulous solution was absolutely new! (See *Second Solution* to Problem 1.5).

A week later, on May 4, 1984, Russel received first prize, which included a $1,000 scholarship and a pocket computer. He brought a surprise for me too: before the award presentation ceremony he handed a piece of paper to me and said "How could I make such a mistake!" The paper contained two different perfect solutions of problem 4, the only problem that Russel did not solve completely during the Olympiad!

First winner receiving first prize. Left to right: Dean of Engineering James Tracey, Dean of Education Dennis Mithaug, winner Russel Shaffer, Alexander Soifer, Robert Hall, and Jerry Klemm

At that time I knew Russel well: he was taking my junior college course. Russel combined genuine talent with extreme modesty. After the Olympiad he told me that he had been accepted by the Massachusetts Institute of Technology (MIT) and was not sure whether he was good enough to do well there. I assured him that no place, even MIT, could get enough students like him. "You will be among the very best at any place," I told him.

Russel graduated from MIT with a perfect 5.0 grade average, and went on to earn a doctoral degree in theoretical computer science from Princeton University supported by a most prestigious National Science Foundation (NSF) Scholarship. He worked as a researcher for MCI [and as of the time of the new 2010 edition, works as an independent consultant]. I wonder how many Olympiads Russel would have won if he had not been a senior when the first Olympiad was held.

Sharing second prize honors were juniors John Williams, Palmer High School; Martin Puryear, Cheyenne Mountain High School; Chris Ford, Air Academy High School; and sophomore Scott Johnson, Mitchell High School. Each of them was awarded a $250 scholarship and a scientific calculator.

Four contestants received first honorable mention, and 20 received second honorable mention.

Participants of the Olympiad came from Denver and Castle Rock, Calhan and Ellicot, Woodland Park and Pueblo, and of course, Colorado Springs. Many of them came back a week later for the award presentation ceremony. We wanted to use this forum for something meaningful in addition to presenting awards. I gave a lecture *"How Coloring Can Solve Mathematical Problems."* Thus, we started a tradition of offering lectures on enjoyable topics of mathematics at the award presentation ceremonies.

The Olympiad was born. Now many knew what it was, including us, the judges of the Olympiad. Following is an assessment of the Olympiad sent by one of the judges, a Hewlett-Packard engineer, to her manager:

April 30, 1984

To Tom Saponas:

The first annual Math Olympiad at UCCS on Friday, April 27, was a great success. Approximately 200 junior and senior high school students from the area participated in a well-organized and challenging math contest. Dr. Alexander Soifer and students from his math courses at UCCS did an outstanding job organizing the event.

Dr. Soifer feels that academic excellence does not receive the attention it should. He pointed out that every day you can pick up the newspaper and read about the exploits of high school athletes. An outstanding mathematician, on the other hand, receives no such notice. In addition to the annual Science Fair and the local math contests, he feels the Olympiad is one of the few opportunities our community has to recognize math excellence and to encourage enjoyment and enthusiasm for the subject.

Although my participation this year was very limited, I hope to be able to participate in a more meaningful way next year. HP's participation was sincerely appreciated. The two HP calculators are excellent prizes. I sincerely agree with Dr. Soifer's feelings regarding the need to encourage and recognize math skills, and I have made a personal commitment to participate more next year. I would like to request that HP review their level of participation also. The first annual Math Olympiad was a success and I feel that with increased participation from individuals and local companies this event will improve yearly. Dr. Soifer and the students in his UCCS classes have successfully initiated a worthwhile event and need our support.

Helen Muterspaugh

Problems 1

Problem 1.2 was contributed by the visiting German professor Jörg Stelzer. My childhood friend Semjon Slobodnik and I created Problem 1.5 in 1971, when we both were still undergraduate students. The magazine *Kvant* published it and paid us a whopping (by Russian standards of the time) $40, which allowed Semjon and I to visit bars and restaurants several times. This was the favorite of all problems I created [until 2003, when I created my new favorite problem, *Chess* 7×7, that appeared as Problem 21.5 in the 21st Colorado Mathematical Olympiad in 2004 - see it in a future book]. I adapted other problems from Russian mathematical folklore.

1.1 A tailor with a real fancy for cutting cloth has ten pieces of material. He decides to cut some of these pieces into ten pieces each. He then cuts some of the resulting pieces into ten pieces each. He then cuts some of these resulting pieces into ten pieces each. He continues this way until he finally tires and stops. He proceeds to count the total number of pieces of cloth he now has; after a few minutes work he determines this number to be 1984. Prove that his count must have been incorrect.

1.2 (*J. Stelzer*). We start with n nodes, where n is a positive integer. Each pair of nodes is connected by a single wire. Show that there is always a way of placing a diode on each wire so that the resulting electrical system contains no current loops. (Remark: a diode allows electricity to flow in exactly one direction.)

1.3 A hiker got out of his tent, walked 10 miles south, then 10 miles west, then 10 miles north, and ... found himself in front of his tent! Find *all* points on earth where the tent could have been pitched.

1.4 For his 100th birthday party George invited 202 guests. They presented him with a rectangular birthday cake, with, of course, 100 candles on it. (No three candles, or two candles and a corner of the cake, or one candle and two corners of the cake were on the same straight line.) George cut the cake into triangular pieces by straight cuts connecting candles with each other and/or with corners of the cake, without any cut crossing previously made cuts (and every candle used in a cut). Prove that there are enough pieces to serve each guest a piece of cake, but none will be left for George.

1.5 (A. *Soifer and S. Slobodnik*). Forty-one rooks are placed on a 10×10 chessboard. Prove that you can choose five of them that do not attack each other. (We say that two rooks "attack" each other if they are in the same row or column of the chessboard.)

Solutions 1

1.1. Every time a piece of cloth is cut into 10 pieces, the total number of pieces increases by 9. Therefore, in order to end up with 1984 pieces we must have an equality $1984 = 10 + 9n$ for some integer n, i.e.,

$$1974 = 9n.$$

In the last equality, however, the right side is divisible by 9, whereas the left side is not. This contradiction proves that the tailor's count was incorrect. He could not have possibly ended up with 1984 pieces. ∎

1.2. Let us assign each node an integer, so that all integers $1, 2, \ldots, n$ are used (and, therefore, used once). For a wire connecting nodes i and $j, i < j$, we place a diode to allow the flow from the node i to the node j. As you can see, in our electrical system electricity flows only in the direction of increasing numbers, therefore, the system has no loops. ∎

1.3. Of course, the tent could have been pitched at the North Pole: going 10 miles south down a meridian, then 10 miles west along a parallel, and 10 miles up another meridian would bring you back to the North Pole (See Figure 1.1).

There are, however, other solutions! Imagine what would happen if 10 miles down a meridian were to bring you to a parallel (near the South Pole) that is 10 miles in circumference (Figure 1.1). Yes, you would cover the entire parallel while walking 10 miles west. You would then back track your way along the same meridian, and end up at your tent! Of course, we do not get just a point A: all points of that parallel would be right for the tent. And not only them!

You may go down a meridian to a parallel that is $\frac{10}{2}$ miles in circumference, and your starting point would work.

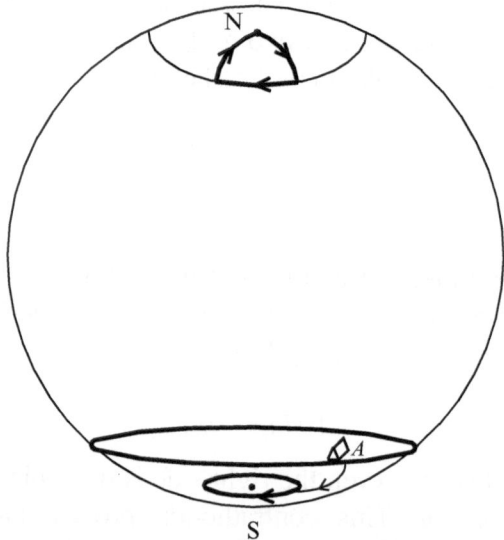

Fig. 1.1

The same is true for parallels $\frac{10}{n}$ miles long in circumference for any positive integer n. Thus, the tent could have been pitched at the North Pole or on one of the *infinitely many parallels* that are located 10 miles north from the parallels of circumference $\frac{10}{n}$ for any positive integer n. It is easy to verify (do) that there are no other solutions. ■

1.4. I will present two solutions of this problem.

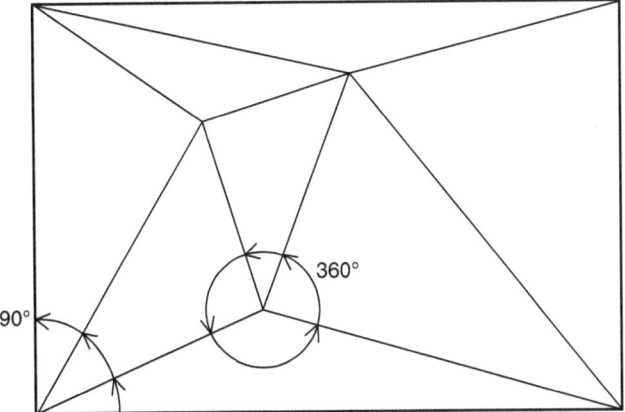

Fig. 1.2

First Solution. Assume that George cut the cake into n triangular pieces in accordance with all conditions of the problem (Figure 1.2). We can calculate the total sum S of angles in the n triangles as follows: we get 360° around every of the 100 candles, plus four 90° corners of the cake, i.e., 360° × 100 + 90° × 4. On the other hand, the sum of angles in a triangle is 180°, therefore, $S = 180° \times n$. By solving the equation

$$180° \cdot n = 360° \cdot 100 + 90° \cdot 4$$

we get $n = 202$. Please note that the *number of pieces does not depend upon a particular way (and there are many ways) in which we cut the cake!* ∎

Second Solution. This solution requires knowledge of the following famous Euler's Formula:

For any map in the plane,

$$F + V = E + 1,$$

where F, V, and E are the numbers of countries, vertices and boundaries respectively.

   In our case $V = 104$ (100 candles plus 4 corners of the cake). Also, since every "country" (pieces of cake are our "countries"!) is *triangular*, $3F$ would be equal to the total number of boundaries. Every boundary, except for the four sides of the cake, separates *two* "countries", therefore we counted them twice in $3F$. Hence, we get the following relationship:

$$3F = 2E - 4,$$

i.e., $E = \frac{3}{2}F + 2$. Substituting $\frac{3}{2}F + 2$ for E and 104 for V in Euler's Formula, we get

$$F + 104 = \frac{3}{2}F + 2 + 1.$$

Hence, $F = 202$. ∎

1.5. I have got to present three solutions of this problem, for they all are so beautiful!

First Solution, by the authors of the problem, A. Soifer and S. Slobodnik.

(a) Since 41 rooks (pigeons) are placed in ten rows of the board (pigeonholes), and $41 = 4 \times 10 + 1$, by the Pigeonhole Principle, there exists a row A with at least five rooks on it.

   If we remove row A, we will have nine rows (pigeonholes) left with at least $41 - 10 = 31$ rooks (pigeons) on them. Since $31 > 3 \times 9 + 1$, by the Pigeonhole Principle, there is a row B among the nine rows with at least four rooks on it.

   Now we remove rows A and B. We are left with eight rows (pigeonholes) and at least $41 - 2 \times 10 = 21$ rooks (pigeons) on them. Since $21 > 2 \times 8 + 1$, by the Pigeonhole Principle, there is a row C among the eight rows with at least three rooks on it.

   Continuing this reasoning, we get row D with at least two rooks on it, and row E with at least one rook on it.

(b) Now we are ready to select the required five rooks.

   First we pick any rook R_1 from row E (at least one exists there, remember!).

   Next we pick a rook R_2 from row D that is not in the same column as R_1. This can be done too, because at least two rooks exist in row D.

   Next, of course, we pick a rook R_3 from row C that is not in the same column as R_1 and R_2. Conveniently, it can be done since row C contains at least three rooks.

Continuing this construction, we end up with the five rooks R_1, $R_2, \ldots, R_5$ such that they are from different rows (one per row, out of the selected rows, A, B, C, D, E) and from different columns, therefore, they do not attack each other. ■

Second Solution, first found during the 1984 Olympiad (!) by Russel Shaffer; the idea of using symmetry of all colors by gluing a cylinder, belongs to my university student Bob Wood.

Let us make a cylinder out of the chessboard by gluing together two opposite sides of the board, and color the cylinder diagonally in 10 colors (see Figure 1.3).

Now we have $41 = 4 \times 10 + 1$ pigeons (rooks) in 10 pigeonholes (one-color diagonals), therefore, by the Pigeonhole Principle, there is at least one hole that contains at least 5 pigeons. But the 5 rooks located on the same one-color diagonal are not attacking each other! ■

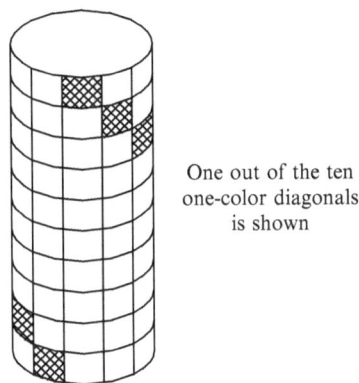

One out of the ten
one-color diagonals
is shown

Fig. 1.3

Third Solution, found by the Digital Equipment Corporation software engineer George Berry right after the Olympiad, when he read the problem in the local newspaper *Gazette Telegraph.* His solution was contained in the first e-mail I ever received in my life.

Suppose that you could choose at most four rooks that are not attacking each other. Select any four rooks that are not attacking each other. We will call these four the key rooks.

Now consider an operation on the chessboard that exchanges two rows or two columns of the board. This exchange does not affect the "attacks" of any of the rooks.

We can use this to "normalize" the chessboard, that is to put the four key rooks into locations $(0,0)$, $(1,1)$, $(2,2)$, and $(3,3)$, where the first coordinate determines the row, and the second coordinate the column of a rook (please see Figure 1.4).

The normalization procedure can be done as follows: locate any one key rook and exchange its row with row 0 and its column with column 0; do likewise for the other three key rooks (moving them into $(1,1)$, etc.). At the end of normalization the chessboard looks like the board shown in Figure 1.4.

	0	1	2	3	4	5	6	7	8	9	
0	K	?	?	?	?	?	?	?	?	?	
1	?	K	?	?	?	?	?	?	?	?	— = there is
2	?	?	K	?	?	?	?	?	?	?	no rook
3	?	?	?	K	?	?	?	?	?	?	
4	?	?	?	?	—	—	—	—	—	—	K = occupied
5	?	?	?	?	—	—	—	—	—	—	by key rook
6	?	?	?	?	—	—	—	—	—	—	? = don't
7	?	?	?	?	—	—	—	—	—	—	know whether
8	?	?	?	?	—	—	—	—	—	—	there is a rook
9	?	?	?	?	—	—	—	—	—	—	there

Fig. 1.4

Consider the square of the chessboard with coordinates (n,m) or (m,n) such that $0 \le n \le 3$ and $4 \le m \le 9$. There are 48 such squares. We will call these the *outside squares* of the board.

Select a pair of outside squares with coordinates (n,m) and (m,n). Is it possible for both of these squares to contain a rook? No, for if both squares had rooks in them, then we could select a new set of key rooks by taking rooks on those two squares along with three of the four key rooks [omitting the one at (n,n)]. This would form a new set of *five* key rooks: we eliminated the only one of the original key rooks that attacked (n,m) and (m,n); and obviously (n,m) does not attack (m,n). However a set of *five* key rooks is impossible by the original assumption that the largest possible set of key rooks was four.

Therefore, at most one of the pair of outside squares (n, m) and (m, n) is occupied. That means that at most there are 24 rooks on the outside squares (half the squares are occupied). When added to the 16 rooks (at most on the *inside* squares), we have a total of 40 rooks accounted for. That is one less than the 41 rooks given in the problem, so the assumption that there are at most four key rooks must be false. ∎

For generalization of these ideas, please see Chapter E1 of *Further Explorations* in this book.

Second Colorado Mathematical Olympiad
April 19, 1985

Historical Notes 2

The First Olympiad was such a success, it received such wonderful media coverage, that I was called into the office of my dean. He went straight to the point:

"You ran the Olympiad, I solicited the prizes. What do we need the College of Education for?"

"You want to join Education as a co-organizer?" I asked.

"No, Engineering alone will run the Olympiad from now on."

"I cannot go to Dennis (Dean of Education), the only person who supported the Olympiad from the start, and tell him to get out."

"I'll do it myself," said my dean.

A few days later my Dean's secretary, Gloria Lawlis, informed me that the issue had been resolved and the College of Engineering was to run the Olympiad alone. Shortly after that I talked to the Education Dean Dennis Mithaug, at our new Chancellor Neal Lane's home party. I told him how disappointed I was with my dean's action. Dear Dennis replied that his job of supporting my idea of the Olympiad was accomplished, that the Olympiad was born, and he was glad that my dean was now willing to support it. The Olympiad had thus landed in the College of Engineering (but not for long: see *Historical Notes* 3 for the continuation of this story).

Chancellor Neal Lane enthusiastically endorsed the Olympiad, and spoke at the Award Presentation. Moreover, in a conversation with me he envisioned every spring, the time of rejuvenating nature, as the time to celebrate young talents not only in mathematics, but also in

A. Soifer, *The Colorado Mathematical Olympiad and Further Explorations:*
From the Mountains of Colorado to the Peaks of Mathematics,
DOI 10.1007/978-0-387-75472-7_4, © Alexander Soifer, 2011

physics, biology, and chemistry. On February 25, 1985, he wrote a letter to Physics Chair Jim Burkhart (and copied me in):

> I spoke with Alex Soifer this morning about the Mathematics Olympiad he runs at UCCS. What is the chance that we might organize a Physics Olympiad or a joint Math and Physics Olympiad along the same lines?

My natural sciences colleagues did not seem to wish to invest considerable time and energy that the Olympiad undertaking would have required.

At the beginning of every Olympiad I walk through each of some 15-20 rooms and answer questions of contestants

Problem 2.4 of the Second Olympiad was contributed by Professor Steven Janke of Colorado College. All other problems I adapted from the Russian mathematical folklore.

Three very important sponsors came on board this year. John Wicklund, one of the teachers of the first winner Russel Shaffer, and Anne Thrasher, the district's mathematics coordinator, were instrumental in bringing about the Colorado Springs School District #11's sponsorship. The new wonderful Chancellor of UCCS Neal Lane started a tradition of sponsorship of the Olympiad by the Chancellor's office. The City of Colorado Springs started donating the City memorabilia.

In the First Olympiad Richard Wolniewicz received first honorable mention. This time the Air Academy High School senior from the class of Judy Williamson won first prize, which included a $600 scholarship and a computer by Radio Shack. In addition to a perfect score for the first four problems, he was the only one to substantially advance in Problem 5.

Richard Wolniewicz receiving first prize. Left to right: Chancellor Neal Lane, Alexander Soifer, Bob Rennick of Digital, and Richard

Just three years later, in 1988, Richard graduated from the University of Colorado at Boulder with a degree in Engineering Physics. Two years after that he earned a master's degree in Computer Science. As I was writing the 1994 version of this book, on February 16, 1994, Richard Wolniewicz received his doctorate in Computer Science from the University of Colorado. Together with two friends, Richard was planning to start a software company. I hope it will become another American success story, challenging the famous Microsoft.

Second prize was awarded to a junior high school (!) student, a ninth grader, David Hunter, who received a $400 scholarship and books on mathematics donated by several judges. He submitted perfect solutions for the first four problems. Do not forget his name: you will see it in the next three chapters as well as in the Part V of this book!

Three students were awarded third prizes: Tim Wood, a senior at Air Academy; Christopher Pounds, a Wasson junior; and Irving ninth-grader Bill Beltz were each awarded a Hewlett-Packard calculator and University and City memorabilia.

Fourth-prize awards went to 6 Olympians; we also awarded 17 first and 11 second honorable mentions. The City newspaper, the *Gazette Telegraph*, which had usually provided wonderful coverage of the Olympiad, this time ran only the list of first through fourth prize winners. I sent the editors of the *Gazette* the following challenge:

> The *Gazette* listed our first-fourth prize winners in " Senior takes Top Math Honor" (May 1, p. B7). They apparently decided that first or second honorable mentions of the Olympiad are not high enough achievements to be mentioned in the local paper. I am taking this opportunity to challenge the Gazette editors to enroll in the Third Annual Colorado Springs Mathematical Olympiad (to be held in April, 1986) and see for themselves whether they will win an honorable mention! I will forward to them the exact date of the event as it becomes known.
>
> Having participated myself in about 30 Olympiads, I have great admiration for the 28 youngsters who competed in the field of 231 fine scholars and won honorable mentions. I hope their classmates, teachers, schools, and neighbors will join me in applauding them.

The editors did not accept the challenge, but on May 8, 1985, all names of the winners of honorable mentions were published in *The Gazette*. Moreover, the Editor-in-chief, Tom Mullen, assured me that "names of the Olympiad winners are information, and the Gazette is in the business of publishing information." He offered to meet with me, aa news editor, and a journalist every year a month prior to the Olympiad to discuss and plan their coverage.

The award presentation featured a good number of speakers (please see some of them presenting first prize to Richard Wolniewicz on a photograph earlier in this chapter, and Councilman Frank Parisi in *Historical Notes 4*) and a lecture *Fibonacci and the Golden Ratio* by Gene Abrams, an assistant professor at UCCS.

Problems 2

2.1 A box contains 100 marbles. There are 30 red ones, 30 blue ones, 30 green ones; the remaining 10 consist of some black ones and some white ones. If we choose marbles from the box without looking, what is the smallest number of marbles we must pick in order to be absolutely certain that among the chosen marbles at least 10 are of the same color?

2.2 Twenty-five basketball teams are entered in a tournament that will last ten days. Show that at the end of the fourth day of the tournament at least one of the teams has played an even number of games. (Remark: zero is an even number.)

2.3 Each of the 49 entries of a square 7×7 table is filled by an integer between 1 and 7, so that each column contains all of the integers 1, 2, 3, 4, 5, 6, 7 and the table is symmetric with respect to its diagonal D going from its upper left corner to its lower right corner. Prove that the diagonal D has all of the integers 1, 2, 3, 4, 5, 6, 7 on it.

2.4 (*S. Janke*). Suppose you are given a list consisting of the numbers 1, 2, 3, 4, 5, 6, 7, 8, 9 in some order, where each number occurs exactly once. You try to put the list in increasing order using the following procedure. Compare the numbers in the first and second positions. If they are in increasing order, make no changes; if they are not in increasing order, switch them. (If you made a switch, the number originally in the first position is now in the second position.) Now compare the numbers in the second and third positions. As before, switch them if they are not in increasing order. Continue in this fashion, finally comparing numbers in positions eight and nine, switching if necessary. As the example shows, this procedure does not always put the list in increasing order.

> *Example*:   Original list   : 2 1 3 4 9 6 5 7 8
> After procedure : 1 2 3 4 6 5 7 8 9

How many different original lists are put in increasing order after using our procedure?

2.5. Is it possible to put 1985 straight line segments in the plane in such a way that each segment has each of its endpoints lying on an inside point of some other segment?

Solutions 2

2.1. Clearly, 37 marbles are not enough: we can get 10 blacks and whites and 9 each of red, blue, and green.

On the other hand, 38 marbles do guarantee that we have at least 28 marbles that are neither black nor white. Since $28 = 3.9 + 1$, by the Pigeonhole Principle we are certain to have at least 10 marbles of the same color. ■

2.2. Let G be the total number of games played by the end of the fourth day, and $g_1, g_2, \ldots, g_{25}$ the number of games played by the end of the fourth day by the first, second, ..., the twenty-fifth team respectively. Since a game between, say, teams i and j contributes one unit each to g_i and g_j, i.e., a total of 2 units, we get the following equality:

$$g_1 + g_2 + \ldots + g_{25} = 2G.$$

Since the number on the right side is even, and the number of summands (25) on the left side is odd, we can conclude that at least one of the summands g_i must be even. ■

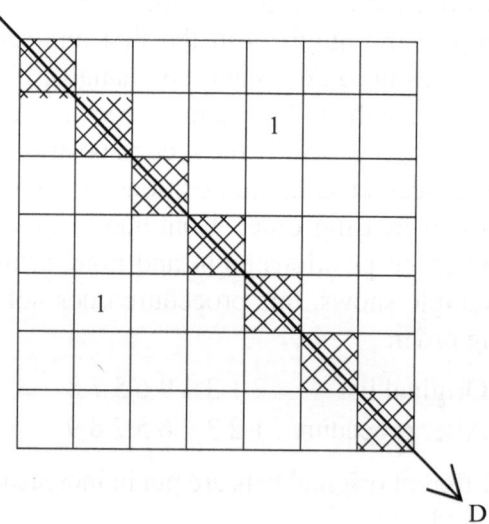

Fig. 2.1

2.3. Due to the given symmetry, for every integer 1 off the diagonal, there is another integer 1, symmetric to the first, which also lies off the diagonal (Figure 2.1).

Therefore, the number of integers 1 off the diagonal is even. The total number of integers 1 in the table is odd (one per each of the seven columns). Thus, at least one integer 1 must be on the diagonal.

Similarly, the diagonal must contain the integers 2, 3, 4, 5, 6, and 7. ■

2.4. Note that *our procedure can move a number any amount of spaces to the right, whereas it can move it only one space to the left.*

Since after the use of our procedure, we want 1 to be in position one, 2 in position two, etc., we have the following options for positions of numbers in the original list.

Number 1 can be in position one or two	2 options
Number 2 can be in position one, two, or three, but not in the position occupied by 1	2 options
Number 3 can be in positions one through four, but not in two positions occupied by 1 and 2.	2 options
. . . .	. . . .
Number 8 can be in position one through nine, but not in the seven positions occupied by $1, 2, \ldots, 7$	2 options
Number 9 can be in the only position remaining after $1, 2, \ldots, 8$ are placed.	1 option

Thus, there are $2 \times 2 \times 2 \times 2 \times 2 \times 2 \times 2 \times 2 = 2^8 = 256$ different original lists that are put in increasing order by our procedure. ■

2.5. Assume that there is a system S of 1985 line segments in the plane such that each endpoint of each segment is an inside point of some other segment of S.

The projection of a point a on the line L is the base a' of the perpendicular drawn from a to L. The projection of a segment is the union of the projections of all its points (see Figure 2.2).

Since there are only finitely many (1985 to be exact) segments in S, we can choose a line L that is not perpendicular to any segment of S. Let us now project all segments of S on L and denote the union

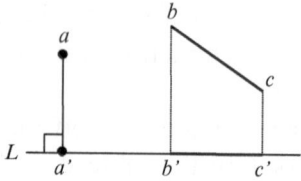

Fig. 2.2

of all these projections by S'. The first point p' of S' from the right (Figure 2.3) must be the projection of an endpoint p of one of the segments of S (can you prove this?).

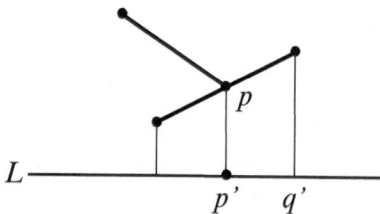

Fig. 2.3

Since the endpoint p must be an inside point of another segment, there will be a point q' in S' to the right of p'. This contradiction to p' being the first point of S' from the right, proves that such a system of segments does not exist.

Please note: 1985 has nothing to do with the result we proved. Such a system of segments does not exist for any nonzero number of segments. ∎

Third Colorado Mathematical Olympiad
April 18, 1986

Historical Notes 3

As you recall, Dean Tracey declined to host the First Olympiad, insisting on hosting the Second Olympiad. He probably had the idea of parity at heart, for on October 21, 1985, he called me in to say how little he cared about the Third Olympiad. I replied with the following October 22, 1985 letter to him:

> You mentioned yesterday that the Olympiad has been a low priority for the College of Engineering and Applied Science (EAS) and that is the reason why I visited you yesterday to ask for secretarial and fund raising assistance. I have a different opinion. The College of EAS was the sponsor of the Olympiad in 1985 because of your choice; and I visited you yesterday to find out whether you would like to sponsor it again in 1986 and treat it as a high enough priority for smooth operation of this large event.
>
> I believe that the Colorado Springs Mathematical Olympiad is by now a very important part of Pikes Peak Region life, well deserving to be a high priority of its sponsors (it was a high priority of the School of Education in 1984).
>
> I also believe that this Olympiad delivers an incredible amount of publicity and visibility for the college, campus and the University for an amazingly low cost.
>
> I think, therefore, that not only Olympiad but also UCCS would lose, if we were to move the Olympiad to a dedicated sponsor outside of the campus.

A. Soifer, *The Colorado Mathematical Olympiad and Further Explorations: From the Mountains of Colorado to the Peaks of Mathematics*, DOI 10.1007/978-0-387-75472-7_5, © Alexander Soifer, 2011

In view of the above, I would like to know:

(1) Whether the College of Engineering would like to sponsor the Olympiad and provide all the necessary secretarial and some financial support.

(2) Whether you would not mind to continue your personal contribution to the Olympiad in raising the prize fund.

Shortly thereafter, Dean Tracey and my chair of the Mathematics Department, Rangaswami, called me on the dean's carpet.

"We are not in a position to help the community. We have to concentrate on our own business: research, curriculum development. You should stop doing the Olympiad," said the Dean, opening the meeting.

"I see: the community helped you get the new Engineering Building and is helping now the college with obtaining a Ph.D. program, but we are not in a position to help the community," replied I, "OK, I will start looking for a new host."

"No, **You** should not continue doing the Olympiad."

"I understand: you no longer wish to be a host. You are certainly in the position to stop the College's patronage, but why should **I** stop doing the Olympiad?"

"One can make an argument in your tenure evaluation, that you do it at the expense of acceptable research and teaching."

"Do you have a problem with my research?"

"No."

"Do you have a problem with my teaching?"

"No, you are a very popular teacher."

"Well, then what I do in my free time is none of your business!"

As I was leaving the room, I heard Rangaswami telling the dean, "I am Alex's friend, but I agree with you." These words brought to my mind the Old Russian proverbial prayer: *Save us, God, from our friends. We will deal ourselves with our enemies.* Dean Tracey and his "willing executioners" (to borrow the term from Daniel Goldhagen's book on the Third Reich) did attempt (without eventual success!) to use the Olympiad against me in the tenure evaluation that was already on the way. In a year, Rangaswami and fellow professor Gene Abrams set the record of hypocrisy: in secret from me, they used the Olympiad's reputation and my Olympiad problems in their application to get money from the Colorado Commission on Higher Education to replace me by a paid director to run the Olympiad! Read this episode in the *Historical Notes 5.*

As to Dean James H. Tracey, he set a world record of anti-intellectualism by limiting to 2 the number of bookcases in a professor office ("you have too many books," he told me), and treating book writing as the lowliest priority in the college. His June 28, 1883 memo entitled "Book Writing" opened as follows:

> I've learned over the past year that we have a considerable amount of interest among the faculty in book writing.

One would expect the dean to praise and support such an "interest." Instead, the dean ended his memo with the promise to treat writing books as the "lowest priority" in the college:

> ... Beginning Fall Semester, 1983, book writing will be handled just as consulting. Book writing will be subject to the same rules and regulations as defined under "Additional Remuneration for Consultation Services" in the Faculty Handbook. Release time will fall under the 1/6 rule and staff will not be available for assistance except for a modest amount of editing or on a lowest priority basis.

If you were a diligent reader, then you would easily guess where I went straight from this infamous meeting with the dean and the chair. Yes, I went to the Dean of Education Dennis Mithaug. Dennis did not just accept the return of the "prodigal" Olympiad. With great enthusiasm, he went into organizing our meetings with Dr. Thomas Crawford (Superintendent of the Colorado Springs District 20), Dr. Harold Terry (Superintendent of the Harrison School District 2) and many other superintendents of school districts. For the first time I was able to tell five superintendents that *this was their Olympiad, this was their students' Olympiad.*

The response was overwhelming: Colorado Springs District 11 dramatically increased its contribution. Air Academy District 20, Harrison District 2, Lewis Palmer District 38, Cheyenne Mountain District 2, and Widefield District 3 became sponsors. Now they took much better care of informing their students of the forthcoming Olympiads. Participation jumped to 405 students (more than the first two years combined), representing 17 districts of the state and a number of private schools. Prize funds increased to surpass the first two years combined as well.

Problems 3.1, 3.3, and 3.5 of this year I adapted from the Russian mathematical folklore. Problem 3.4 was a particular case of a problem I have used in the motivational opening of my then forthcoming book *How Does One Cut a Triangle?* [S2] (see also the new expanded edition [S13]). Problem 3.5 was the simplest observation related to the open problem of the chromatic number of the plane (see *Further Explorations*, chapter E2 later in this book).

And the winner was David Hunter, *already a sophomore* at Palmer High School in the class of Victor Kuklin. He received first prize: a scholarship for a three-week International Summer Institute in South Hampton, Long Island, New York. There David attended lectures by famous scientists, Nobel Laureates, and my daily problem solving course. Among his 4 solved problems ($4\frac{1}{2}$, to be precise) Dave's was the only solution of Problem 3.5. This brought him a special prize for creativity sponsored annually by Robert Penkhus, a Volvo-Mazda dealer.

Three second prize winners each received a $300 scholarship and a scientific calculator by Hewlett-Packard. They were Ronald Wright from Manitou High School; David Coufal from Mitchell High School; and Rhett Rodewald from St. Mary's School. Ron and Rhet also participated in the International Summer Institute in South Hampton.

Three students won third prize: Gideon Yaffe, The Colorado Springs School; Scott Johnson, Mitchell; and Pete Rauch, Air Academy. They each received a Hewlett-Packard calculator.

We also awarded 13 fourth prizes, 25 first honorable mentions, and 69 second honorable mentions.

This year's award presentation unveiled the Olympiad logo and poster that were created especially for us by the great artist (and my father) Yuri Soifer. We used his design to produce the Olympiad posters; gold, silver, and bronze medals for the winners of the top three prizes of the Olympiad; and the buttons for the winners of honorable mentions, as well as for proctors and judges of the Olympiad.

On June 17, 1991, Yuri Soifer lost battle with cancer, but his medals continue to decorate the brightest young minds of Colorado. In 2009

I dedicated to Yuri, "who introduced colors into my life" my most important work, *The Mathematical Coloring Book* [S11], which took me 18 years to write.

Problems 3

3.1 The first digit of a six-digit number N is equal to the fourth digit, the second is equal to the fifth and the third is equal to the sixth. Prove that N is evenly divisible by 7, 11, and 13.

3.2 Santa Claus and his elves paint the plane in two colors, red and green. Prove that the plane contains two points of the same color exactly one mile apart.

3.3 A finite number of arcs of a given circle are painted red. The total length of the red arcs is less than half of the circumference of the circle. Prove that there exists a diameter such that neither of its endpoints is on red.

3.4 (*A. Soifer*) A rectangular pool table of size $p \times 2q$, where p and q are given odd integers, has pockets in every corner and in the middle of each side of length $2q$. A ball is rolled from a corner under a $45°$ angle with respect to the rails. Prove that the ball will get into a pocket; moreover, this will be a side pocket. (The angle of incidence is equal to the angle of reflection.)

3.5 Given n integers, prove that either one of them is a multiple of n, or a number of them add up to a multiple of n.

Solutions 3

3.1. Let the bar above the digits denote the decimal representation of a number. We have:

$$\overline{abcabc} = 1001 \cdot \overline{abc} = 7 \cdot 11 \cdot 13 \cdot \overline{abc}. \quad \blacksquare$$

3.2. Toss on the plane an equilateral triangle with side lengths equal to one mile. Since its three vertices (pigeons) are painted in two colors (pigeonholes), there are two vertices painted in the same color (at least two pigeons in a hole). These two vertices *are* one mile apart. ■

This simple problem leads to exciting ideas and my favorite open problem of mathematics. Please find it in Chapter E2 of *Further Explorations* in this book.

3.3. Denote by S the union of all red arcs, and by S' the result of rotation of S through 180° about the center of the circle. Since the union of S and S' does not cover the circle completely (do you see why?), there is a point p on the circle that lies neither in S nor in S'.

Neither endpoint of the diameter through p is red (verify this). ■

If you liked this problem, look for more problems continuing these ideas in Chapter E3 of *Further Explorations*.

3.4. Every time the ball hits a rail, we can reflect the pool table symmetrically with respect to that rail. Since the angle of incidence is equal to the angle of reflection, the trajectory of the ball reflected a number of times will become a straight line! (Please see Figures 3.1, 3.2, 3.3, and 3.4).

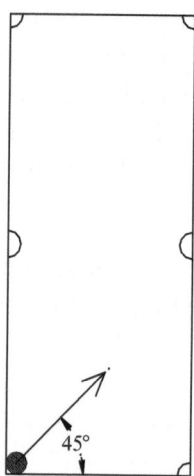

Fig. 3.1

As a result, our problem becomes equivalent to the following:

Prove that a straight line drawn from the origin at a 45° angle with respect to the coordinate axes will pass through a point with coordinates (mp, nq) for some positive integers m, n (see Figure 3.5).

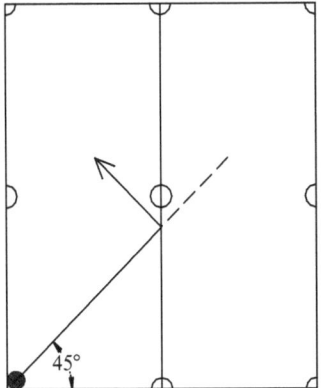

Fig. 3.2

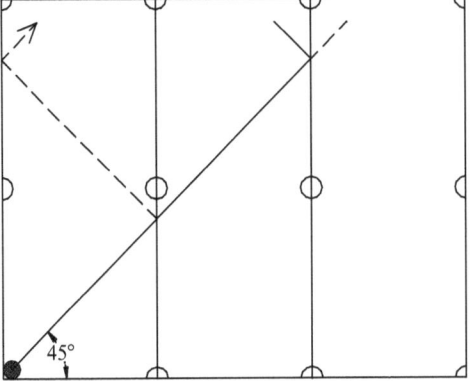

Fig. 3.3

Since the equation of our straight line is $y = x$, it will pass through the point (mp, nq) if and only if

$$mp = nq. \tag{*}$$

If we choose $m = q$ and $n = p$ the equality (*) will be satisfied. Thus, the ball will get into a pocket.

Will this be a side pocket or a corner pocket? To answer this question, we need to determine which dots of the lattice in Figure 3.5 represent side pockets and which represent corner ones. It is easy to notice that the side pockets are represented by the dots (mp, nq) precisely when n is *odd*.

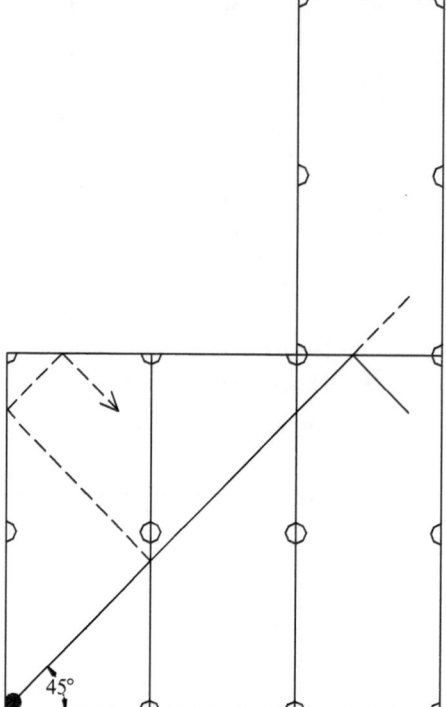

Fig. 3.4

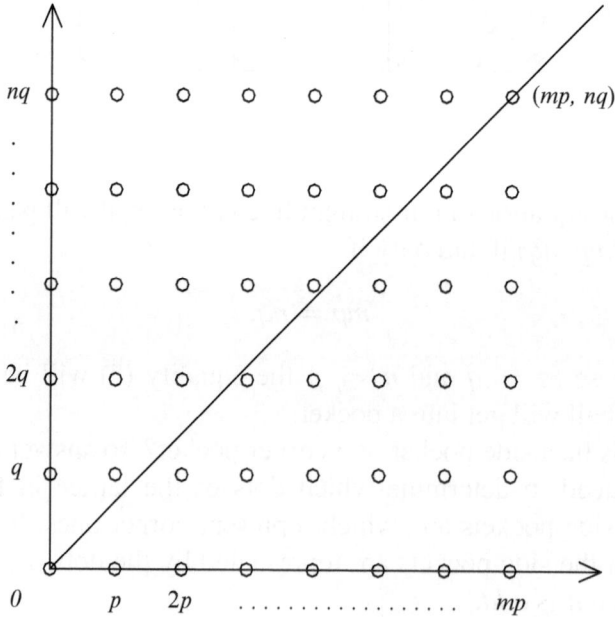

Fig. 3.5

The first pocket on the way of the ball corresponds to the smallest positive solution for m and n of the equation (*). Since p and q are odd, the smallest positive m and n that satisfy the equation (*) will be odd as well. Hence the ball will get into a side pocket. ∎

3.5. Denote the given integers by $a_1, a_2, \ldots, a_n$. Define:

$$S_1 = a_1$$
$$S_2 = a_1 + a_2$$
$$\vdots$$
$$\vdots$$
$$S_n = a_1 + a_2 + \ldots + a_n.$$

If one of the numbers $S_1, S_2, \ldots, S_n$ is a multiple of n, we are done. Otherwise all possible remainders upon division of these numbers by n are $1, 2, \ldots, n - 1$, i.e., we get more numbers S_i (namely n, *they* are our "pigeons") than possible remainders ($n - 1$ possible remainders, they are our "pigeonholes"). Therefore, among the numbers $S_1, S_2, \ldots, S_n$ there are two numbers, say S_k and S_{k+t} which give the same remainder upon division by n.

We are done because

(1) $S_{k+t} - S_k$ is a multiple on n;
(2) $S_{k+1} - S_k = a_{k+1} + a_{k+2} + \ldots + a_{k+t}$.

In other words, we have found a set of the given numbers, namely $a_{k+1}, a_{k+2}, \ldots, a_{k+t}$, whose sum is a multiple of n. ∎

Fourth Colorado Mathematical Olympiad
April 17, 1987

Historical Notes 4

In early fall 1986 I received a surprising call from Ms. Bailey Barash, an anchorwoman at CNN, whom I saw daily on my TV screen. A CNN news team was to come to Colorado Springs and cover an important international gathering at the Broadmoor Hotel. A CNN education crew wanted to come along and cover the Olympiad, if it were to be "served" in November. I did not feel we could make such a move, even for CNN. On CNN's request, I reminded them in March about their interest in covering the Olympiad, but, of course, without a major political event in the neighborhood they *were* interested in the Olympiad, but not *too* interested.

Paul Zeitz joined the judges for the first time during the Third Olympiad, when he worked as a mathematics teacher at the Colorado Springs School. It was a natural thing to do for a former winner of USA and International Mathematical Olympiads. It was extraordinary that he participated in the Fourth Olympiad. By this time Paul was a mathematics doctoral student at the University of California, Berkeley. Only his passion for Olympiads brought Paul back to Colorado Springs for one weekend of grading!

I was sure that nobody in Colorado knew the very nice problem created by Mikhail Serov and used in the *Fourth* National Soviet Union Olympiad in 1970 (Mikhail and I both were among the 30 judges of that Olympiad). I decided to use it for our *Fourth* Olympiad. It became Problem 4.5.

A. Soifer, *The Colorado Mathematical Olympiad and Further Explorations: From the Mountains of Colorado to the Peaks of Mathematics*, DOI 10.1007/978-0-387-75472-7_6, © Alexander Soifer, 2011

Problems 4.1, 4.2, and 4.3 were adapted from the Russian mathematical folklore. I created Problem 4.4 for this Olympiad. In fact, my lecture *"How Does One Cut a Triangle?"* at the award presentation took problem 4.4 on an exciting tour of generalizations. You will find some of these ideas in chapter E4 of *Further Explorations* later in this book, and all of them in my book [S2] and its new edition [S13].

There were 432 students from 18 school districts and a number of private schools who participated in the Olympiad. For the first time we used Yuri Soifer's design to produce the Olympiad's gold, silver, and bronze medals for the winners of first through third prizes, and Olympiad buttons to go with honorable mentions.

This year we had the first tie for the victory. First prizes were awarded to David Hunter (his second victory in a row!), now a junior in Victor Kuklin's class at Palmer High School, and to Michael Eamon, a junior in Judy Williamson's class at Air Academy High School. Each of them received the gold medal, a $2,000 scholarship for the International Summer Institute, and a Hewlett-Packard calculator.

Second prizes were awarded to Debbie Allison, a senior from Longmont, and to Patrick Harlan from Rampart High School. They received silver medals, calculators, and books.

Third prizes were awarded to Gideon Yaffe from Colorado Springs School (remember him, for we will meet him again in the next Historical Notes and in Part V), Michael Lanker from Doherty High School, and three students from the class of Mr. Proctor at Woodland Park High School: Erik Wiener, Paul Simmons and Mark Ruth. These and other youngsters from the mountain town of Woodland Park showed great perseverance. They were seventh graders when they first came to the Olympiad. Their third participation was a charm!

New administrators came to UCCS. Chancellor Dwayne C. Nuzum and Dean of Education Margaret A. Bacon for the first time spoke at the award presentation ceremony. Both of them became staunch supporters of the Olympiad, major contributors to its success. Robert Penkhus, the sponsor of the special prize for creativity, awarded it to David Hunter. Once again we were honored to have an address by Frank Parisi, a member of the Colorado Springs City Council.

City of Colorado Springs Councilman Frank Parisi addressing the winners

Problems 4

4.1 Given three barrels, an 8 liter barrel full of water, an empty 5 liter barrel, and an empty 3 liter barrel, describe the process of the consecutive pouring of water from barrel to barrel that ends up with 4 liters of water in the 8 liter barrel and 4 liters of water in the 5 liter barrel. (The barrels have no measuring marks on them, and thus all you can do is either empty a barrel completely into another one or fill a barrel to its capacity.)

4.2 1987 points are plotted in the plane; no three points lie on the same straight line. The points are connected by 1987 line segments such that each given point is an endpoint of exactly two segments. Prove that no straight line can be drawn in the plane that passes through inside points of each of the 1987 segments. (An inside point of a segment is any segment's point that is not its endpoint.)

4.3 Each square of a chessboard that is infinite in every direction contains a positive integer. The integer in each square equals the mean of the four integers contained in the squares that lie directly above, below, left, and right of it. Show that every square of the board contains the same integer.

4.4 (*A. Soifer*).

(A) Give an example of a triangle that cannot be cut into two triangles similar to each other.
(B) On the other hand, show that any triangle can be cut into six triangles similar to each other.

4.5 (*M. Serov*). Each side of an equilateral triangle is partitioned into n segments of equal length, and straight lines parallel to the sides are drawn through the points of the partition. As a result, we get n^2 *little triangles* congruent to each other. A sequence of little triangles is called a *chain* if no triangle appears more than once in it, and every triangle beginning with the second one has a common side with the previous triangle of the sequence. What is the maximum number of little triangles a chain can have?

Solutions 4

4.1. It is convenient to represent the state of water affairs by an ordered triple (a, b, c), where a, b, and c are the amounts of water (in liters) in the 8 liter barrel, the 5 liter barrel, and the 3 liter barrel respectively. The initial state is (8,0,0). One of the sequences that solves the problem is the following:

$$(8, 0, 0) \to (3, 5, 0) \to (3, 2, 3) \to (6, 2, 0) \to$$
$$(6, 0, 2) \to (1, 5, 2) \to (1, 4, 3) \to (4, 4, 0). \quad \blacksquare$$

4.2. Assume that there is a line L that passes through inside points of each of the 1987 segments. It is easy to verify (do) that L does not pass through any of the given points. The line L divides the plane into two half-planes S_1 and S_2.

Pick one of the given points in S_1, call it p_1. The point p_1 is an endpoint of one of the given segments that intersects L. Therefore,

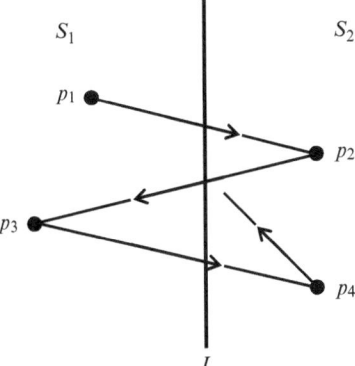

Fig. 4.1

the other endpoint p_2 of this segment lies in S_2. One more given segment originates in p_2; since it intersects L, the other endpoint p_3 of this segment lies in S_1. One more given segment originates in p_3; since it intersects L, the other endpoint p_4 of this segment lies in S_2 (Figure 4.1).

We continue this process until one of the given points is repeated (and thus a loop created). This point must be p_1, because for all other points along our path we have already accounted for two segments originating at them, and only two given segments are allowed to start at a given point by the conditions of the problem. So far we have accounted for an equal number of given points in S_1 and S_2, i.e., for an even number of given points total. Since the number of the given points is odd (it is 1987), we have a given point not yet counted. Call it g_1. We start with g_1 using exactly the same process as with p_1 above: we go to g_2, to g_3, etc., until g_1 is repeated. By then we have accounted for an even number of given points.

Therefore, there is a given point, say r_1, not yet counted. And so on. This process of counting given points will never stop, which is a contradiction to the finite number of the given points. Thus, such a line L does not exist. ■

4.3. Let us call two integers on a chessboard *neighbors* if they are contained in squares of the board that share a side. Assume that not all squares of the board contain the same integer. Then there are two neighbors a_1 and a_2 that are not equal: $a_1 > a_2$. Since a_2 is the average of its four neighbors, one of which, namely a_1, is greater than a_2, there is a neighbor a_3 of a_2 that is smaller than a_2. We have $a_1 > a_2 > a_3$.

This process can be continued indefinitely, producing an infinite decreasing sequence of positive integers

$$a_1 > a_2 > a_3 > \ldots > a_n > \ldots,$$

which is absurd; therefore, every square of the board contains the same integer.

For the continuation of these ideas, try to solve Problem 10.5 on your own. ■

4.4 (A). Assume that a triangle ABC is cut into two triangles similar to each other. A cut must go through a vertex of the triangle ABC and reach the opposite side (otherwise we will not get two triangles). Let BD be this cut (Figure 4.2).

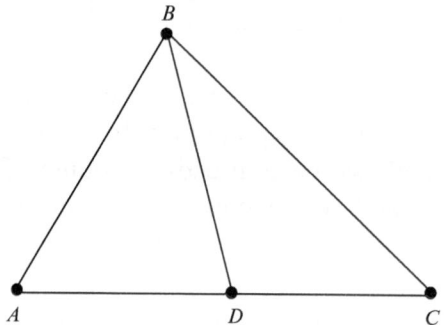

Fig. 4.2

Let $m(\alpha)$ denote the measure of an angle α. Without loss of generality we can assume that $m(\angle BDC) \geq m(\angle BDA)$. Since also $m(\angle BDC) = m(\angle DAB) + m(\angle ABD)$ and the triangles ABD and BCD are similar, we conclude that the following equality must hold:

$$m(\angle BDC) = m(\angle BDA) = 90°.$$

Moreover, the similarity of triangles ABD and BCD implies that $m(\angle BAD) = m(\angle BCD)$ and thus the triangle ABC is *isosceles*; or else $m(\angle BAD) = m(\angle DBC)$ and the triangle ABC is a right triangle [do you see why in the latter case $m(\angle ABC) = 90°$?].

Thus, a nonisosceles nonright triangle *ABC* cannot be cut into two triangles similar to each other.

(B) Divide an arbitrary triangle into nine congruent triangles by lines parallel to its sides, and combine the upper four triangles into one (see Figure 4.3). ∎

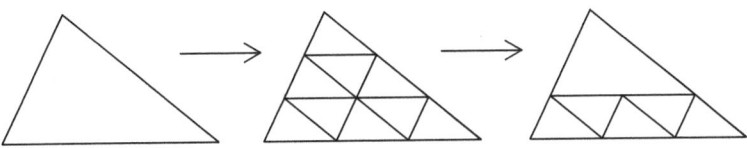

Fig. 4.3

For much more general related problems, please see Chapter E4 of *Further Explorations*.

4.5. At first glance, all n^2 little triangles appear alike. Nothing can be further from the truth! Some little triangles have a vertex above their horizontal base while others have not. To emphasize this, we color little triangles of the first type black, and leave the rest white (please, see Figure 4.4).

Coloring for $n = 10$

Fig. 4.4

Observe: *in any chain black and white triangles alternate*! Since there are n more black triangles than white ones (one extra black triangle in each of the n rows), a chain may not have more than $n^2 - n + 1$ triangles (1 is added because a chain may start and finish with a black triangle).

On the other hand, there is a chain, the "snake chain," that contains exactly $n^2 - n + 1$ triangles (see it in Figure 4.5). ∎

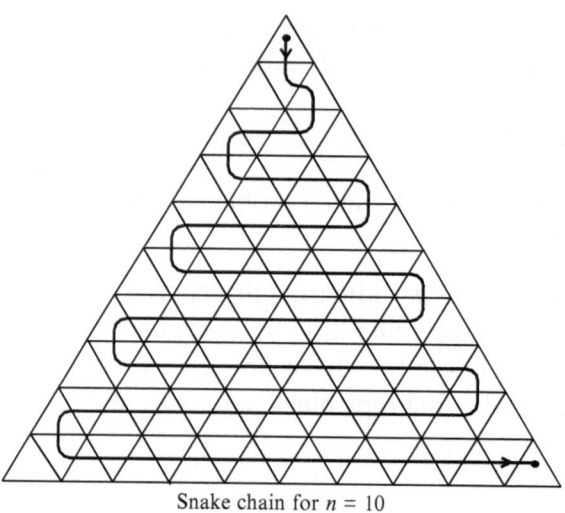

Snake chain for $n = 10$

Fig. 4.5

Fifth Colorado Mathematical Olympiad
April 22, 1988

Historical Notes 5

As you recall from Historical Notes 3, Dean Tracey and my Chair, Rangaswami, ordered me to close down the Olympiad, and threatened with consequences to my employment if I did not. In 1987 Rangaswami, using the draft by another perpetrator in my tenure process, Professor Gene Abrams, produced "Proposal to the CCHE (The Colorado Commission on Higher Education) for the Promotion and Encouragement of Excellence." Rangaswami asked for funding of the "Center of Resources for Excellence in and Advancement of Mathematics (CREAM)." He wrote:

> The Olympiad, which has been held each April since 1984, attracts hundreds of junior high and high school students to the campus. They spend four hours answering a series of five questions, each of which requires mathematical insight and creativity to answer correctly (Appendix 4).[1] In this way, two impressions are left on the students: both that mathematics is really more than just tedious memorization, and (perhaps more importantly) that knowledge and intelligence are virtues to be rewarded and applauded. In addition students become familiar with the UCCS campus, and thus will hopefully become more inclined to make use of campus resources (e.g. using the Microcomputer Lab, eventually enrolling as a student, etc.).

---

[1] The Appendix contained all the Olympiad problems to date (which in itself constituted unauthorized use of copyrighted material).

A. Soifer, *The Colorado Mathematical Olympiad and Further Explorations:*
From the Mountains of Colorado to the Peaks of Mathematics,
DOI 10.1007/978-0-387-75472-7_7, © Alexander Soifer, 2011

And this was the same chair who, together with the dean, demanded that I close down the Olympiad! This proposal was kept in secret from me, and it was due only to negligence of the conspirators that I noticed it lying near the copy machine in the math department office. You must be wondering, why was it a secret? Because, the canning chair and Professor Gene Abrams wanted to get the money from the State of Colorado to take the Olympiad away from me: to hire a paid Director reporting to the chair, who among other duties would be in charge of the Olympiad. The proposal reads:

> The director will be responsible for coordinating the UCCS [sic] Mathematical Olympiad.

My inquiry Chair Rangaswami defiantly rebuked with "This is not your Olympiad!" He was more conciliatory in May 1987 when Dean of Education Dennis Mithaug asked him to remove any mention of the Olympiad from the proposal. Rangaswami promised, but violated his word to Mithaug, and did not answer my May 29, 1987 inquiry into the state of this attempted theft of the Olympiad. On October 21, 1987, I took the matter to the Vice Chancellor for Academic Affairs, Joan Klingel:

> Last May (!) Dr. Rangaswami promised me and Dr. Mithaug to either change his CCHE "CREAM" proposal in the parts where it deals with the Colorado Springs Mathematical Olympiad (CSMO) and present it to me for approval, or to drop any mention of CSMO from the proposal. It is October 21, 1987 (!), and he still has not done so. Moreover, I received no response to my May 29, 1987 memo (attached).
>
> Hereby I am giving you a chance to resolve the matter internally to my satisfaction. If it is not done soon, I will have no alternatives, but to advise CCHE on the record of Math. Department and the College of EAS [Engineering and Applied Science] in trying to shot down CSMO, and their total lack of support for CSMO.

The Vice Chancellor's reply came the very next day. She "explained" Rangaswami's violations by CCHE's cancellation of the fund, but assured me that

> If funds are ever appropriated again to a Centers of Excellence Competition, and if Professor Rangaswami decides to submit a proposal for Math Department on [sic] the style of CREAM he will contact you before including the Math Olympiad in any such proposal.

One would not expect such an opposition to a clearly worthy event such as the Olympiad. However, lust for power and control are in the nature of some administrators, who will do anything to be "in charge." They always prefer subordination to academic freedom, and a cowardly all-against-one approach.

I have experienced exodus twice in my life. Both were due to my search for freedom, first of all human rights and academic freedom. The first exodus happened when I left the Soviet Union on June 21, 1978. The second exodus took place on January 1, 1988. I "emigrated" from the totalitarian hills of the College of Engineering to "the land of the free," the College of Letters, Arts, and Sciences.

My new dean, James A. Null, deeply surprised me. "I want you to do in research, teaching, and service what *you* want to do," he began our first meeting. Having noticed my disbelief, Dean Null added: "This way you will do your best for our College. Your achievements *are* our achievements. If you are not burned out, we would love you to continue your Olympiad. It is a great contribution to the community and the University." From this December 1987 conversation for the duration of his deanship James Null has been a loyal supporter of the Olympiad and a friend always ready to help in my numerous creative projects. He never ceased to greet me with "Hi, professor. What can I do to help?"

As in 1984 and 1985, a question of the host for the Olympiad came up. Unlike my former Engineering dean, however, James Null told me, to my delight, that "even though we can run the Olympiad alone, I would much prefer to run it jointly with the College of Education. There is an important symbolism in the College of Letters, Arts, and Sciences and the School of Education hosting the Olympiad together." This is how the Olympiad has worked since then for many years.

Dean James A. Null, College of Letters, Arts and Sciences

At the same time, in early January 1988, Chancellor Dwayne C. Nuzum has made the Olympiad an official function of the University by creating and funding a permanent Olympiad line in the budget of the College of Letters, Arts, and Sciences. Thank you, Dwayne!

These positive events were accompanied with great pre-Olympiad publicity. For the first time the Olympiad appeared on the pages of Cañon City's *Daily Record* (April 11, 1988), and it was the first article on the front page! *The Gazette Telegraph* ran a very nice story about Erik Wiener by Raymond McCaffrey (April 18, 1988) on the front page of the News section. Here is an excerpt from it:

> Some teen-agers run from problems. Erik Wiener courts them.
>
> If some teacher hasn't posed a dilemma such as the one above, Wiener will create his own.
>
> "Creativity is essential to it," said the 16-year old Woodland Park junior. "That's what I like . . . just the fact that you have to use your creativity to solve a problem."

Wiener will have the perfect opportunity to use his creativity Friday at the Fifth Annual Colorado Springs Mathematical Olympiad. Last year, he placed third in the contest.

"This year I have to do better than third place," Wiener said. "I've sort of set a hard standard for myself."

The Olympiad, to be held at the University of Colorado at Colorado Springs, will draw some pretty tough competitors. They will include Palmer High School senior David Hunter, who has placed first in the Olympiad the past two years. If Hunter wins this year, it will be an unprecedented feat.

"It's a record that may withstand the century," said UCCS Professor Alexander Soifer, who helps organize the Olympiad.

Wiener is just one member of a strong Woodland Park delegation that has grown up with the Olympiad. Wiener was in the seventh grade when he first competed three years ago. "We didn't really prepare for it," he said "We didn't know what to expect."

But the Woodland Park contingent has improved through the years. Last year, three students from the small Teller County city tied for third in the Olympiad.

Five days later, another enjoyable article by Ray McCaffrey appeared on the front page of the News section in *The Gazette Telegraph*. This one covered the Olympiad still in progress. As a result, Ray's article preserved something intangible, the spirit of the participants:

The teen-agers lay contentedly in the thick grass, enjoying the Friday afternoon sunshine and an endless wave of music pulsing from a radio.

Alexander Soifer eyed the group of high school students as he strode across the University of Colorado at Colorado Springs campus. The UCCS Professor couldn't help but notice it wasn't yet 1 p.m. – the official completion time for most of the estimated 650 participants in the fifth annual Colorado Springs Mathematical Olympiad.

"I predict that those who are winners will not submit paper to proctor until time is up," Soifer said , his voice laced with a heavy Russian accent. "David Hunter is still working."

The news that Hunter, a Palmer High School senior and two-time first-place finisher in the Olympiad, was still at work came as a shock to the lounging students. Some members of the group finished the five question test by 11 a.m.

"Why would he still be working?" asked Liberty High School freshman Dawn Boden.

Classmate Andrew Pedersen knew the answer to Boden's question was a question – five to be exact.

"They were all pretty hard," Pedersen admitted.

Even Hunter – who finished second as a freshman, won outright as a sophomore, and shared top honors with Air Academy High School Junior Michael Eamon last year – found the Olympiad test difficult.

"It was rough," Hunter said. "Probably rougher than last year ... I didn't have time to double check."

The difficulty of the test can best be seen by one of the problems Hunter found easiest: . . .

The remaining four questions on the test were solely developed by Soifer, who started the Olympiad. Soifer himself competed in similar competitions while a student in his native Moscow.

"You have to pass the baton to the next generation," Soifer said.

Soifer said 170 students came to the first Mathematical Olympiad. Last year, 432 students competed in the contest.

"I thought it stalled at 400," Soifer said. "Now we have 200 more."

Of the many contestants who grew up with the Olympiad, none is more renowned than Hunter.

"His consistency is something phenomenal," Soifer said.

Hunter was well aware he is expected to do well once again. The winners will be announced at an awards ceremony Friday at UCCS.

"When you've gotten a second and two firsts you always feel some pressure to do well," Hunter said. "But I try not to put too much emphasis on winning."

Math is not Hunter's only hobby. He played soccer for Palmer and has been a member of the student cabinet.

"I hate the idea people have of the stereotypical math nerd," Hunter said. "Because I don't see myself that way at all."

Nonetheless, Hunter has found few who share his passion for math. He hopes that will change next year when he is a freshman at Princeton University.

"It's kind of frustrating not being able to share the solutions with someone," Hunter said. "I'm hoping to find some of that at Princeton."

Even those not fanatical about math enjoyed coming to the Olympiad.

"For the experience," said Liberty High's Andrew Pederson.

"Plus," added classmate Colin Wilson, "we got to miss a day of school."

David Hunter from the class of John Barber at Palmer High School did win an unprecedented third first prize! Combined with one second prize (when he was a freshman), this does constitute a record that in my opinion may withstand the century, but we will see! :-)

Before leaving for Princeton University, Dave shared with me his plans. He wanted to earn an undergraduate degree in mathematics, then "pay back to the system," as he said, "by teaching in a high school, and then go to graduate school." David implemented his noble plans. In 1992 he graduated from Princeton, and for nearly two years has been teaching in Hudson, New Hampshire. Coming Fall David will start the graduate school in statistics at the University of Michigan, Ann Arbor. [Added in the 2011 edition: David is now a tenured Professor of Statistics at Pennsylvania State University; read his reminiscences in Part V.]

Three time winner of the Olympiad David Hunter. *Photograph © by Stegner Portraits, Inc.*

New that year at the Olympiad was a sequence of closely related problems of increasing difficulty: 4(A) to 4(B) to 5(A) and 5(B) created by me. This was meant as an illustration of mathematical research, of how one solved problem gives birth to the next one, and so on, creating a train of mathematical thought. This made the Fifth Olympiad harder than its predecessors. On the other hand, it inspired participants to explore new frontiers.

We awarded one more first prize this year. It went to Gideon Yaffe from the class of Bruce Hamilton at Colorado Springs School, who earned second honorable mention in 1985, and third prizes in 1986 and 1987. In solving Problem 4(B) Gideon came up with an idea that inspired me to find a far reaching generalization of Problem 5.5 (please see *Further Explorations,* chapter E4, in this book). Gideon started at Harvard University as a mathematics major, then moved to mathematically intensive areas of philosophy, and finally graduated from Harvard with a degree in philosophy. For a year he worked as an assistant editor of *Mac World* magazine and in 1993 entered the graduate school of Stanford University in philosophy. [Added in the 2011 edition: Gideon is now a double Professor of Philosophy and Law at the University of Southern California; see his letter in Part V.]

First prize winners, David and Gideon, each were awarded the gold medal and a $1,000 scholarship.

Erik Olson, an Air Academy junior, and Chris Hall, a sophomore at Cheyenne Mountain, shared second prize. Each of them received the silver medal, a $500 scholarship, and a Hewlett- Packard calculator.

Third prize winners were Michael Eamon, an Air Academy senior who shared first prize the year before, and Matt Kahle, an Air Academy freshman [you will meet Matt again in *Historical Notes* for 1990 and 1991, and read his letter at the end of Part V.]

In 1992 Mike Eamon earned a Computer Science degree from The University of Colorado at Boulder, and has since been a principal software engineer for "Analytic Spectral Devises."

Seventeen participants were awarded first honorable mention, and forty five students were awarded second honorable mention.

Regular speakers at the award presentation this year were joined by State Senator Michael C. Bird, who found the time to attend in spite of the demanding job of Chairman of the Joint Budget Committee of the State General Assembly; UCCS Chancellor Dwayne Nuzum; and Deans James A. Null and Margaret Bacon.

My lecture in honor of the winners that year was called *"Problems of Mathematical Olympiads: Where Do They Come From?"*

July 27–August 3, 1988, I spent in Budapest at the International Congress on Mathematical Education, a quadrennial gathering of some 3000-4000 mathematicians and math educators. This was my first congress, and it influenced much in my life. First and foremost, I met Paul Erdős, and our friendship commenced immediately. The following March, Paul was visiting me in Colorado Springs, and our collaboration began. His influence was so profound that I changed my research field from Abelian group theory to "Erdősian" combinatorics, geometry and Ramsey theory.

I also met there the Russian geometer Vladimir Boltyanski, who came to Colorado Springs in June 1990, where we jointly created a draft of *Geometric Etudes in Combinatorial Mathematics* [BS]. I also joined the World Federation of National Mathematics Competitions

(WFNMC), and met many colleagues, leaders of their national Olympiads, who since became my friends: Peter J. Taylor (Australia), Petar Kenderov, Jordan Tabov, and Slavi Bilchev (Bulgaria), Francisco Bellot-Rosado (Spain), Maurice Stark (New Caledonia), Pier-Olivier Legrand (French Polinesia), Maria Falk de Losada (Columbia), and many others.

Problems 5

These problems were selected and edited by the Problem Committee: W. Escovitz, M. Robbins, A. Soifer, and R. Wood.

5.1 You are given 80 coins. Seventy-nine of these coins are identical in weight, while one is a heavier counterfeit. Using only an equal arm balance, demonstrate a method for identifying the counterfeit coin using only four weighings.

An equal arm balance is a device composed of two plates suspended from an arm. By placing a set of coins on each plate, one can determine which set of coins has the greater weight, but cannot determine by how much.

5.2 (*A. Soifer*). Prove that one cannot divide a regular pentagon into exactly 1988 pieces, *each of which is a triangle*, by making a series of consecutive cuts.

A *cut* is a straight line segment connecting two vertices. A new vertex is formed at a point where two or more cuts intersect. New cuts may connect original vertices of the pentagon, new vertices, or an original vertex and a new vertex.

5.3 (*A. Soifer*). Mark put one layer of n^2 red and green blocks on an $n \times n$ chessboard; each block covered exactly one square of the board. He put the first layer arbitrarily, then he remembered that his dad had asked him to put blocks in such a way that:

(1) every red block has common faces with an even number of green blocks; and
(2) every green block has common faces with an odd number of red blocks.

Mark decided to put the second layer of blocks on top of the first one, so that the above two conditions were satisfied for all the blocks of the first layer. If, as a result, the conditions were to be satisfied for all the blocks of the second layer, Mark would stop. If not, Mark would put the third layer of blocks so that the conditions are satisfied for all the blocks of the second layer, etc. Is there such a first layer that this process of putting consecutive layers will never stop?

Statement (*): Among any n points located inside or on the sides of a given triangle T of area 1, there are always three points that are vertices of a triangle of area at most $\frac{1}{4}$.

5.4 (*A. Soifer*).

(A) Prove Statement (*) for $n = 9$.
(B) Prove Statement (*) for $n = 7$.

5.5 (*A. Soifer*).

(A) Prove Statement (*) for $n = 5$.
(B) Prove that Statement (*) is false for $n = 4$, regardless of the shape of the triangle T.

Solutions 5

5.1. Divide the 80 coins into three groups of 27, 27, and 26 coins. By weighing two 27 coin sets against each other we will find which of the three groups has a counterfeit coin.

If the weights of the two 27-coin sets are unequal, then the heavier one contains a counterfeit. If they are equal in weight, then the 26 coin set contains a counterfeit. In the latter case, let us add one coin to the set of 26 coins.

We used one weighing and got 27 coins, one of which is counterfeit. Now we divide the coins into three groups of nine coins, and weigh the first group against the second one. This will allow us, as was the case with the first weighing, to identify the group that contains a counterfeit.

After two weighings we are down to nine coins. We divide them into three groups of three coins each and weigh the first group against the second one.

Finally we divide the remaining three coins into three groups of one coin each, and weigh the first group against the second one. ∎

5.2. I will present two solutions for this problem.

First Solution. Assume that a regular pentagon is divided into 1988 triangles. Then the total sum of the angles of all triangles is equal to $1988 \times 180°$. On the other hand, this sum is equal to the sum of angles of the pentagon, which is $540°$, plus $360° \times n$, where n is the number of the new vertices formed inside the pentagon (the sum of angles around any new vertex is $360°$). We get the following equality:

$$1988 \times 180° = 540° + 360° \times n,$$

i.e.,

$$1988 = 3 + 2n.$$

An even number in the left side may not be equal to an odd number in the right side. This contradiction proves that a regular pentagon (for that matter any convex pentagon, since we did not use the regularity at all) may not be cut into 1988 triangles. ∎

Second Solution. Assume that we cut a pentagon into 1988 triangles. Then the total number of sides of triangles is equal to 1988×3. Each side of each triangle, except for the five sides of the original pentagon, was counted twice in 1988×3. Therefore, we get the equality

$$1988 \times 3 = 2E - 5,$$

where E is the total number of sides of triangles. The left side of the equality is even, whereas the right side is odd.

This contradiction proves that a pentagon (even a *non-convex* one, as long as its sides do not intersect each other) may not be cut into 1988 triangles. ∎

5.3. I will present the solution in the style of the alphabet song: A-B-C-D-E.

(A) ***What does determine the colors of blocks in a layer?*** For the blocks of the second layer it is, of course, the colors of blocks in the first layer. For $k \geq 3$, however, the colors of blocks in layer k are determined by the colors of blocks in the preceding two layers: layers $(k - 2)$ and $(k - 1)$.

(B) *We can go back*: Given colors of blocks in layers k and $(k-1)$, we can uniquely reconstruct the colors of blocks in the layer $(k-2)$.

(C) *How many distinct pairs of consecutive layers colored red & green are there?* Surely *finitely* many (the pair of layers is composed of $2n^2$ blocks, each of which has only two options, to be red or green).

(D) Assume now that there is such a first layer that the process of putting consecutive layers will never stop. This means that we have *infinitely* many pairs of consecutive layers. On the other hand, due to (C), we only have finitely many distinct pairs. Therefore, there is a pair of consecutive layers that *repeats*! In other words, there are layers $(k-1)$ and k that are colored identically to layers $(k-1+m)$ and $(k+m)$ respectively, for some positive integer m (please see Figure 5.1).

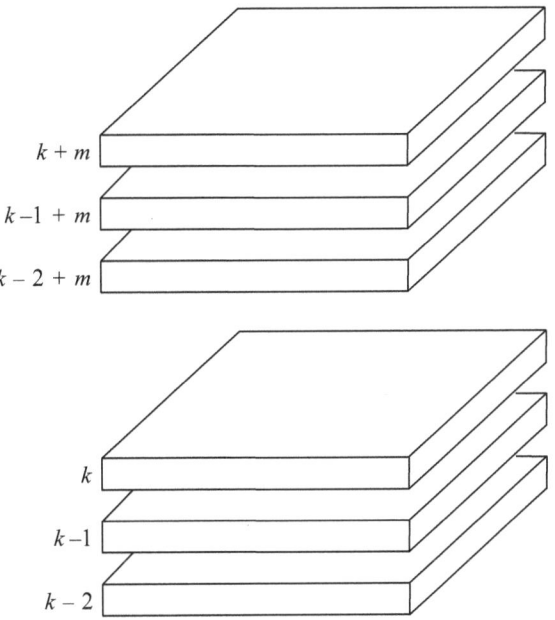

Fig. 5.1

In view of (B), we can go back and conclude that then the layers $(k-2)$ and $(k-2+m)$ are identically colored. Now we have identical pairs of layers $(k-2), (k-1)$ and $(k-2+m), (k-1+m)$. Therefore, layers $(k-3)$ and $(k-3+m)$ are identically colored, and so on. We end up with layers 1 and 2 that are colored identically to layers $(m + 1)$ and $(m + 2)$ respectively. What do we do next? The first layer has no predecessor, thus we cannot go back any more!

(E) Let us go back nevertheless: let us add *layer* 0 under layer 1 that is colored identically to layer 1. Surely, this will not violate the conditions 1) and 2) of the problem for any layer! Since we showed in (D) that the pairs of layers 1, 2 and $(1 + m), (2 + m)$ are identically colored, we conclude, in view of (B), that layers 0 and m are colored identically to each other. Thus layers zero, 1, m and $(m + 1)$ are colored identically. This means that the conditions 1) and 2) were satisfied for the layer m *without* putting layer $(m + 1)$ on! Thus, our process of putting consecutive layers has stopped with the layer m, if not sooner, which contradicts the assumption that started (D). ■

If you enjoyed this problem, see problems 10.4(A) and 10.4(B) and Chapter E9 of *Further Explorations* later in this book.

5.4(A). Midlines partition the given triangle into four congruent triangles of area $\frac{1}{4}$ (Figure 5.2). These congruent triangles are our

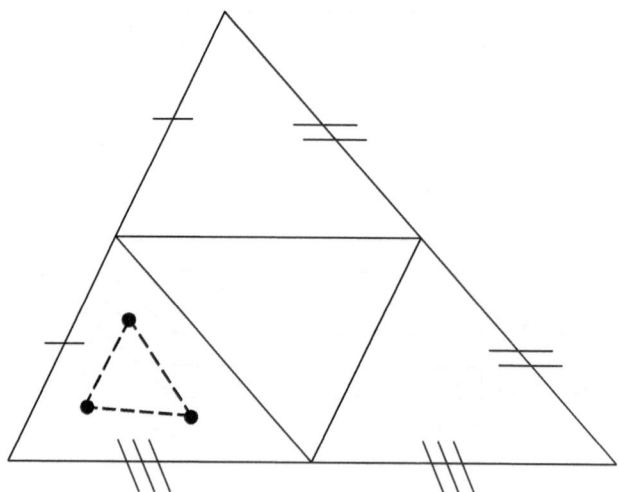

Fig. 5.2

pigeonholes. The given points are our pigeons. Now nine pigeons are sitting in four pigeonholes. Since $9 = 2 \times 4 + 1$, by the Pigeonhole

Principle there is at least one pigeonhole that contains at least three pigeons. In other words, three of the given points are inside a triangle of area $\frac{1}{4}$. Therefore, they form a triangle of area at most $\frac{1}{4}$. ■

5.4(B). Since $7 = 2 \times 3 + 1$, it would be nice to have three pigeonholes: then by the Pigeonhole Principle, at least one of them would contain at least three pigeons! Okay, let us draw only two midlines in the given triangle (Figure 5.3).

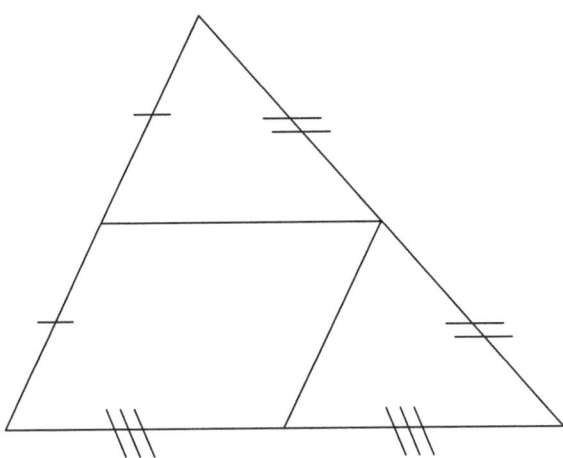

Fig. 5.3

We get three pigeonholes. At least one of them contains at least three pigeons. If one of the triangles contains three given points, we are done.

If the parallelogram contains three given points, then all we have left is to prove a simple tool (I prefer to use the English word "tool," for "lemma" is Greek to me :-).

Tool #1. The maximum area of a triangle inscribed in a parallelogram of area $\frac{1}{2}$ is equal to $\frac{1}{4}$.

The proof of this tool is left to you. ■

5.5(A). The following solution is essentially due to Royce Peng (I just simplified and shortened it). But first another tool:

Tool #2. The locus of all vertices B of triangles ABC with the given segment $\overline{AC}$ as the base, and the given area S is a pair of straight lines

L_1, L_2 parallel to $\overline{AC}$ lying on either side of $\overline{AC}$ at the distance $\frac{2S}{|\overline{AC}|}$, where $|\overline{AC}|$ denotes the length of $\overline{AC}$ (please see Figure 5.4)

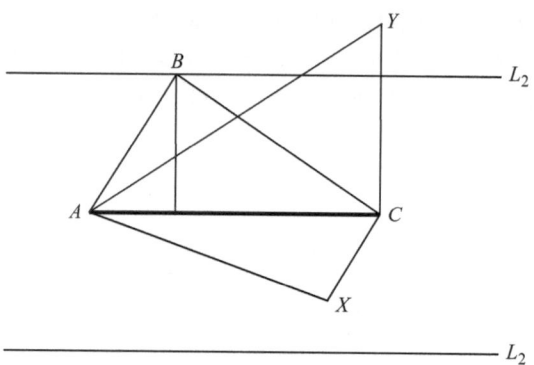

Fig. 5.4

Moreover, if the area of a triangle AXC is less than S, then the point X lies between the lines L_1 and L_2. If the area of a triangle AYC is greater than S, then the point Y lies outside of the strip bounded by the lines L_1 and L_2. I refer readers who need assistance proving Tool #2 or would like to have more problems on loci, to my book *Mathematics as Problem Solving* ([S1], pp. 54-57; or to its new expanded edition [S12]).

Now we are ready for the solution. Midlines (again!) partition the given triangle into four pigeonholes (see Figure 5.4). At least one of the pigeonholes must contain at least two of the five given points. If the midlines triangle MNK contains two given points v_1 and v_2 (Figure 5.5), then we are done.

Indeed, one of the corner triangles, say MBN, must contain at least one of the three remaining given points v_3, and we can surround the three points v_1, v_2, and v_3 by a parallelogram $MBNK$ of area $\frac{1}{2}$. By Tool #1, this guarantees that the area of the triangle $v_1v_2v_3$ is $\frac{1}{4}$ or less.

Assume now that one of the corner triangles, say AMK, contains at least two given points v_1 and v_2 (Figure 5.6).

Now I would like to draw the locus of all vertices X of triangles v_1Xv_2 of area $\frac{1}{4}$. Tool #2 tells us that this locus is a pair of lines L_1 and L_2 parallel to the segment $\overline{v_1v_2}$. Where can L_1 and L_2 lie? Can they intersect MN, for example?

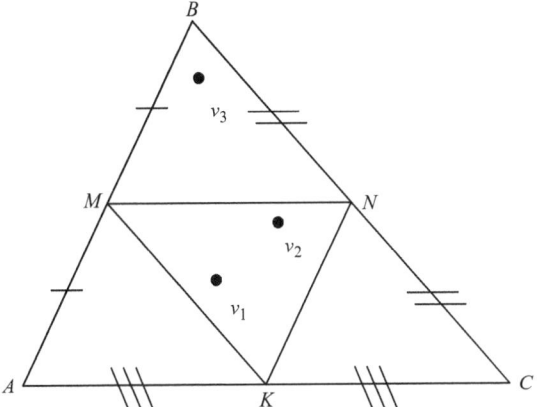

Fig. 5.5

Since each of the triangles $v_1Mv_2, v_1Nv_2, v_1Kv_2,$ *and* v_1Av_2 is contained in the parallelogram $AMNK$ of area $\frac{1}{2}$, by Tool #1, each of these triangles has area at most $\frac{1}{4}$.

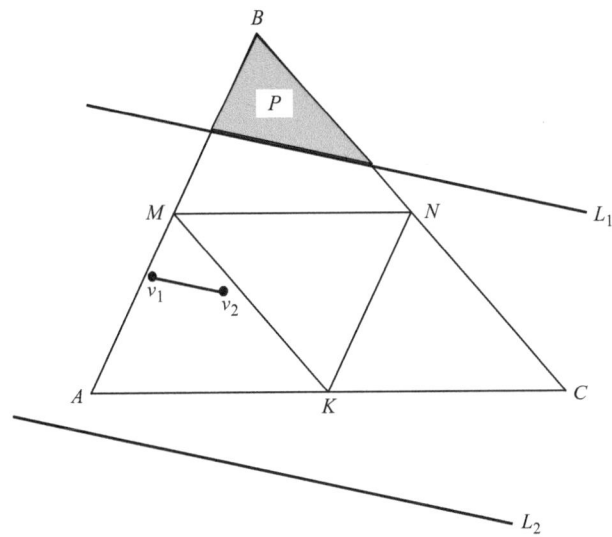

Fig. 5.6

By the second sentence of tool #2, we can conclude now that the points $A, M, N,$ and K all lie between or on the lines L_1 and L_2. What part of the triangle ABC can possibly lie outside of the strip bounded by the lines L_1 and L_2? *Only* a piece P of the triangle *MBN* or of the triangle *NCK*! (Can you prove that pieces of *both* triangles *MBN* and *NCK* may not be outside of the strip?)

Now we are done:

Either one more of the given points v_3 lies between the lines L_1 and L_2 (or on L_1 or L_2), and then the area of the triangle $v_1v_2v_3$ is $\frac{1}{4}$ or less (can you tell why?), or else all three remaining given points $v_3, v_4,$ and v_5 lie in the piece P, and the area of the triangle $v_3v_4v_5$ surely at most equals the area of triangle *MBN*, which is exactly $\frac{1}{4}$!

You can find two more solutions of this problem, both algebraic and elegant (although longer than the one above) in my book *How Does One Cut a Triangle?* ([S2], pp. 53–59, and 131–134; or the expanded edition [S13]). ∎

5.5(B). It suffices to notice that the three vertices of the given triangle plus its center of mass (i.e., the point of intersection of the medians) form the required four points (Figure 5.7). Indeed, the smallest area of a triangle formed by three of the selected four points is equal to $\frac{1}{3}$. ∎

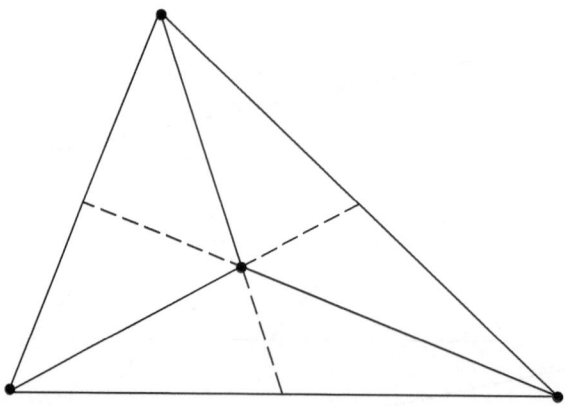

Fig. 5.7

To see where these ideas lead to, open Chapter E5 of *Further Explorations* in this book.

The Sixth Colorado Mathematical Olympiad
April 21, 1989

Historical Notes 6

On June 23, 1988, UCCS Chancellor Dwayne Nuzum offered to the Olympiad a tough, yet prestigious test. He asked me to make a presentation about the Olympiad to the Board of Regents of the University of Colorado (four-campus system). I told the regents what the Olympiad was, how the entire community (education, industry, and government) joined together in this ongoing project and why I founded it. The Olympiad passed the test. When my presentation was over, Chairman (of the Regents) Fowler replied:

> Chancellor Nuzum, thanks for this very exciting program. You know, regents, the gifted and talented (aspect) is what the university is all about; and sometimes you wonder whether, in a free country, people are willing to recognize the child who really does well despite all kinds of adversities – and I think children today have more adversities than we like to admit.
>
> The university *really* does need to do more of these programs, ... put the pressure on these high schools, and the other schools; and it's ... surprising to me how many of those schools really do want it. They want us to provide that kind of intellectual leadership – and we just have to do it.
>
> And I think the regents recognize that there's a spontaneous quality to this kind of leadership from you, Alex, and Jerry Flack, and the other people who are working at these programs. You've done them yourself in the microcosm. If you didn't do it, you know, it wouldn't be done. God Bless You!

A. Soifer, *The Colorado Mathematical Olympiad and Further Explorations: From the Mountains of Colorado to the Peaks of Mathematics*, DOI 10.1007/978-0-387-75472-7_8, © Alexander Soifer, 2011

I had been asked numerous questions about the Olympiad through the years. My dream was to answer all of them by a short documentary film about the Olympiad. To produce a film? Who would fund it? Who would produce it in Colorado Springs? All these problems were solved. Chancellor Nuzum contributed a modest amount x; Dean Null funded the remaining $5x$ of the cost. Richard Olden and his Discovery Video Productions produced a truly wonderful documentary, written and directed by Phil Fredrickson.

Paul Erdős in Colorado Springs. "Two Thinkers," a photo by Alexander Soifer

Due to sheer luck, during that time I was visited by the greatest living mathematician, then 76-years old, Paul Erdős. We worked on some problems, posed other problems.

Paul gave two lectures at the university. He appeared in our Olympiad film. Let me repeat for you Paul Erdős's address to the winners of the Olympiad, captured in the film:

It is important to learn more mathematics. The contest itself wouldn't be that important, but it creates new enthusiasm. From this point of view it is important. Also, it has good effect: it stimulates interest in mathematics...

To all the winners of the Colorado Mathematical Olympiad I wish future successes and I hope they will become mathematicians, preferably pure mathematicians, since I am a pure mathematician, but my second choice would be computer science or various branches of applied mathematics, or possibly physics or engineering.

Paul Erdős also contributed *The Erdős Problem*, one of the harder problems of this year. For the first time the problem editors Gary Miller, Darla Waller and I felt that the problems allowed expressive names. *Football for 23* was contributed by Paul Zeitz, now a Ph.D. student at the University of California, Berkeley. *Tiling Game* and *Mark's Chessboard* were created by me. *Sugar*[3] was inspired by the Russian mathematical folklore and written in the form of a very nice Lewis Carroll-like story by the three problem editors.

The contestants liked our story-problems. They responded with stories of their own. We had no choice but to establish the *Literary Award* of the Olympiad (somebody had to reward literary creativity, when our English Department and the newspaper *The Gazette Telegraph* did not). Our first literary award was presented to Matt Fackelman, a junior from Palmer High School, for a metaphysical argument that mathematics cannot prove anything:

> Metaphysically speaking, this problem can never be proven because, in one's proof he will include only mathematical laws necessary for his proof to work. He disregards the mathematical laws that he feels are not necessary and do not need to be incorporated into his proof. By disregarding these other, unused mathematical laws he will never be completely sure his proof actually proved anything because these unused mathematical laws that he deemed unnecessary could have disproved his proof. Therefore, the mathematician has not proved anything.

It was a thrill for me to receive Matt's letter about his victory and his literary aspirations. "It was the happiest day of my life," he wrote. See his complete letter in Part V, *Winners Speak*.

A record 636 students came to the University to compete in the Sixth Olympiad.

You could indeed learn the geography of the State of Colorado by looking at the school buses parked on campus on April 21, 1989:

Rangely, Blanca, Alamosa, Cañon City, Florissant, Woodland Park, Manitou Springs, Colorado Springs, Calhan, Rush, Peyton, Castle Rock, Larkspur, Parker, Miami-Yoder, Erie, Littleton, Longmont, Fort Collins, Brush.

Judges at work (left to right): Robert Ewell, Keith Mann, Gary Miller, and Alexander Soifer. *Photograph by Mary Kelley, Gazette Telegraph*

First prize was awarded to Pete Rauch, a senior from the class of Judy Williams at Air Academy High School. He solved over $4\frac{1}{2}$ problems out of 5. Pete was the only one out of 636 participants to solve *Football for 23*. He received the gold medal, a $1,000 scholarship, a Hewlett-Packard graphing calculator, my autographed book *Mathematics as Problem Solving*, and City and Olympiad memorabilia. Following his victory, Pete went to the Massachusetts Institute of Technology.

In a 6-way tie, second prizes were awarded to Clark Allred, a senior from Liberty, Scott Burles, a junior form Manitou Springs, Chi-Sun Chui, a senior from Palmer, Eric Nickerson, a senior from Palmer, and Chris Sprague, a junior from Rampart. Each of them received the silver medal, a $250 scholarship, a Hewlett-Packard calculator, my autographed book, and City and Olympiad memorabilia. Special prize for creativity, donated by Bob Penkhus Volvo-Mazda dealership, was awarded to the youngest of the second prize winners, a ninth grader, Mike Rauch, a brother of the first prize winner, Pete Rauch.

In a 5-way tie, third prizes were awarded to Greg Geihsler, a sophomore from Coronado, Dan Hedges, from Irving Junior High, Dennis Hwang, a junior from Palmer, Clay Kunz, a sophomore from Wasson, and Eric Olson, a senior from Air Academy. Each of them received the bronze medal, a Hewlett-Packard scientific calculator, my autographed book, a $25 book certificate, and City and Olympiad memorabilia. The United States Space Command Award, donated by the USSC Chief Scientist Dr. David Finkleman, was awarded to the youngest of the second prize winners, a ninth grader, Dan Hedges.

27 participants were awarded first honorable mention, and 119 students received second honorable mention.

My award presentation lecture this year was called *"Some of My Favorite Olympiad Problems."* It was followed and referred to by a wonderfully funny address by Engineering Manager of the Digital Equipment Corporation Robert Rennick:

> Surprised? I am always surprised when I come to the Math Olympiad, which I have most of the years that it's taken place. Today I was surprised. I learned from the problem right behind us, that you can see better without eyes. And I also found out the real reason why I was never able to easily get out of a bureaucratic institution with a certificate. Mathematics is very important to us at Digital. It is the foundation of our work. We work with it every day. We love it. We think that everyone should have some, and you, obviously, have had a lot. We congratulate you on your participation in the Olympiad, on your achievements, and we wish you the best of luck in your future studies.

Other memorable addresses were given by Colorado State Senator Jeffrey M. Wells, Colorado Springs Councilwoman Mary-Ellen McNally, UCCS Chancellor Dwayne C. Nuzum, UCCS Deans James A. Null and Margaret Bacon, Colorado College Dean David Finley, Past President of the Colorado Council of Teachers of Mathematics John Putnam, and President of Penkhus Motors Robert Penkhus, Jr.

Problems 6

When I came up with the idea of Problem 6.1, my Alaskan student Darla Waller, an Olympiad judge, suggested writing it in the style of Lewis Carroll's *Alice in Wonderland*; Darla and I did so and then Gary Miller, a veteran judge, gave the problem a wonderfully playful and concise title. Problem 6.3 was contributed by Paul Zeitz, and Problem 6.2 by me.

I created Problem 6.4 as an "antidote" to the obsession with coloring solutions while I was a student in Moscow: the problem appeared to require a coloring solution, but the appearance was false.

Paul Erdős's contribution of Problem 6.5 was the highest honor our Olympiad has ever received.

6.1 SUGAR[3] *(A, Soifer, D. Waller, and G. Miller).* One day two little girls were playing together at the breakfast table. Julia, the younger, took eight sugar cubes out of the sugar bowl and stacked them to make a single large cube. She laughed, saying, "Look Olivia, each face of my cube has four perfect squares on it." Not to be outdone, Olivia challenged Julia to color all 24 squares on the surface of the large cube in such a way that any two squares that share an edge are colored in different colors. Julia worked hard, but accomplished the task while using only three colors: red, yellow, and blue. Olivia counted the squares of each color and with the importance of a learned schoolgirl said, "Julia, you colored 9 squares in red, 8 in yellow, and 7 in blue." "I ought to tell your teacher that you can't count," said Julia, and she was right. Why?

6.2 MARK'S CHESSBOARD *(A. Soifer).* Mark has a very special black and white 8×8 chessboard: he can interchange any two rows or any two columns of it. Once he had been interchanging them for a while when I peeked and noticed that the upper left 4×4 quarter of the board consisted entirely of black squares. Can you determine the colors of all the remaining squares of the board? Are they uniquely determined?

6.3 FOOTBALL FOR 23 *(P. Zeitz).* Twenty-three people of positive integral weight decide to play football. They select one person as referee and then split up into two 11-person teams of equal *total* weight. It turns out that no matter who is the referee this can always be done. Prove that all 23 people have equal weight.

6.4 THE TILING GAME (*A. Soifer*). Mark and Julia are playing the following tiling game on a 1988×1989 chessboard. They are alternately putting 1×1 square tiles on the board. After each of them has made exactly 100 moves (and thus they covered 200 squares of the board) a winner is determined as follows: Julia wins if the tiling of the board can be completed by dominoes. Otherwise Mark wins. (Dominoes are 1×2 rectangles, which cover exactly two squares of the board. Tiles may not overlap or stick out of the board's boundary.) Can you find a strategy for one of the players allowing him to win regardless of what the moves of the other player may be? You cannot? Let me help you: Mark goes first! Can you find the strategy now?

6.5 THE ERDŐS PROBLEM (*P. Erdős*). Given a collection of $n+1$ distinct positive integers each less than or equal to $2n$, prove that:

(A) you can always find two relatively prime integers in the collection (two integers are called relatively prime if their greatest common divisor is 1);

(B) you can always find two integers in the collection such that one of them is a factor of the other.

Create examples showing that the statements (A) and (B) can be false if the collection consists of only n integers.

Solutions 6

6.1. Look at a corner of the large cube. Three surface squares meet there, and every two of them share an edge (Figure 6.1).

Therefore, Julia must have used one of each of the three colors on these three squares. The large cube has eight such corners, each with its own adjacent three surface squares. Therefore, Julia had to color 8 squares in red, 8 in yellow, and 8 in blue. ∎

6.2. Observe that interchanging two rows would not change the numbers of black and white squares in any row or in any column. The same is true about interchanging two columns, of course.

The original chessboard had 4 black and 4 white squares in every column and every row. The same must be true about the final board that Mark obtained. This observation allows us to uniquely reconstruct the final coloring of the chessboard (please see Figure 6.2). ∎

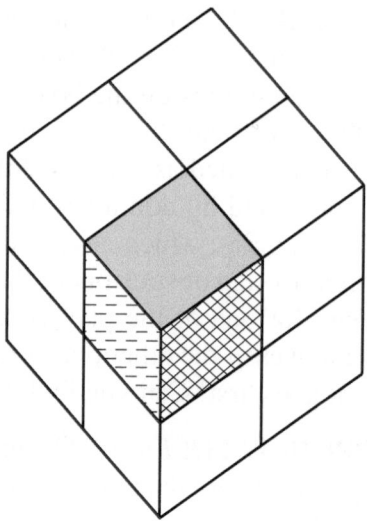

Fig. 6.1

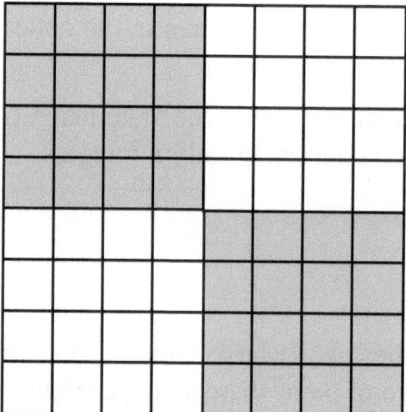

Fig. 6.2

6.3. Assume that there is a set of 23 people satisfying the conditions of the problem (i.e., each can serve as a referee and the rest be split into two teams of equal weight), *and such that not all 23 have equal weight.* Then among such sets of 23 people there is a set with *the smallest total weight.* Let us pick such a set of the smallest possible total weight $A = a_1 + a_2 + \ldots + a_{23}$, where $a_1, a_2, \ldots a_{23}$ are the weights of the people from the set.

Our goal is to find a set of 23 people with smaller total weight than A, and thus reach a contradiction.

First of all, we notice that either all 23 numbers $a_1, a_2, \ldots, a_{23}$ are even, or they are all odd. Indeed, if say, a_1 is even and a_2 is odd, then by choosing a_1 to be referee we see that $A - a_1$ must be even (we divide it into two teams of equal weight, so it must be even). Similarly $A - a_2$ must be even. Therefore, the number $(A - a_1) + (A - a_2)$ is even, but it is equal to $2A - (a_1 + a_2)$ which is odd because $a_1 + a_2$ is odd.

We are now our ready to finish our solution:

(1) If all numbers $a_1, a_2, \ldots, a_{23}$ are even, we replace them by their halves $\frac{a_1}{2}, \frac{a_2}{2}, \ldots, \frac{a_{23}}{2}$, which satisfy all the conditions and have total weight $\frac{A}{2}$, which is less than A.

(2) If all numbers $a_1, a_2, \ldots, a_{23}$ are odd, then we replace them by $\frac{a_1+1}{2}, \frac{a_2+1}{2}, \ldots, \frac{a_{23}+1}{2}$, which satisfy all the conditions and have total weight $\frac{A+23}{2}$, which is less than A. ■

6.4. Julia (i.e., the second player) has a strategy that allows her to win regardless of what Mark's moves may be. All she needs is a bit of home preparation: Julia needs to prepare a tiling template showing one particular way, call it T, of tiling the whole 1988×1989 chessboard. Figure 6.3 gives, for example, such a tiling template T for an 8×13 chessboard.

The strategy for Julia is now clear. As soon as Mark puts a 1×1 tile M on the board, Julia puts her template T on the board to determine which domino D of the template T contains Mark's tile M. Then she puts her 1×1 tile J to cover the second square of the same domino D (Figure 6.4).

After each of the players has made 100 moves (and thus they covered 200 squares of the board), due to Julia's clever strategy, 100 dominoes of her template tiling have been placed. The template shows how to finish the tiling of the board without any problem. ■

6.5(A). We can divide all integers from 1 to $2n$ into n subsets: $\{1, 2\}$; $\{3, 4\}; \ldots; \{2n - 1, 2n\}$. Now we can apply the Pigeonhole Principle: given $n + 1$ integers (pigeons) belonging to (sitting in) n subsets (pigeonholes), there is at least one subset (pigeonhole) that contains at least two integers (pigeons). In other words, for one of the subsets, the

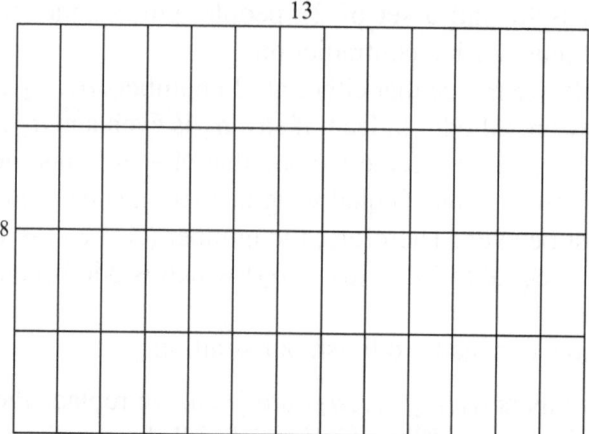

Tiling template T for a 8×13 chessboard

Fig. 6.3

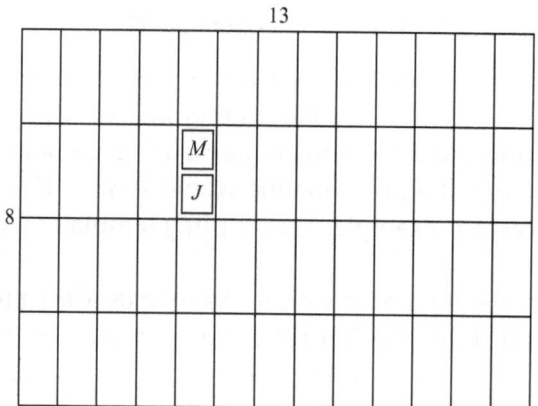

Fig. 6.4

given $n + 1$ integers include both of the integers $k, k + 1$ of that subset. Surely, two consecutive integers k and $k + 1$ are relatively prime.

For n integers the statement of the problem can be false: just take n even integers $2, 4, \ldots, 2n$. No two of them are relatively prime. ∎

6.5(B). We can decompose each of the given $n + 1$ integers $i_1, i_2, \ldots, i_{n+1}$ into two factors, one of which is the maximum power of 2 that is a factor of the given integer, and the other is an odd number (for example, $120 = 2^3.15$):

$$i_1 = 2^{k_1}.m_1$$

$$i_2 = 2^{k_2}.m_2$$

$$\cdots\cdots$$

$$i_{n+1} = 2^{k_{n+1}}.m_{n+1}$$

In these decompositions, all numbers $m_1, m_2, \ldots, m_{n+1}$ are odd and all numbers $k_1, k_2, \ldots, k_{n+1}$ are non-negative integers. Since there are only n odd numbers between 1 and $2n$, two of the $n + 1$ numbers $m_1, m_2, \ldots, m_{n+1}$ are equal. Let us for definiteness assume that $m_1 = m_2$. But then the corresponding integers i_1 and i_2 satisfy the required condition: one is a factor of the other.

To show that n integers are not enough, we can just take integers $n + 1, n + 2, \ldots, 2n$. None of them is a factor of any other. ∎

Seventh Colorado Mathematical Olympiad

April 27, 1990

Historical Notes 7

In August 1989 I taught at the International Summer Institute in Long Island, New York. A fine international contingent of gifted high school students for the first time included a group from the Soviet Union. Some members of this group turned out to be mathematics Olympiad "professionals," winners of the Soviet Union National Mathematics and Physics Olympiads. There was nothing in the Olympiad genre that they did not know. There was no point in teaching these kids problem solving. Instead I offered them – and everyone else – an introduction to certain areas of combinatorial geometry. We quickly reached the forefront of mathematics, full of open problems. I offered my students three $20 open problems that were posed jointly by Paul Erdős and me, and my $50 open problem. The students were ecstatic, and not only they. The leader of the Soviet delegation, a fabulous mathematician and man, Professor Alexei Sosinsky, came to me and said:

"The guys told me that you offered prizes for open problems."

"Yes, three $20 prizes, and one $50 prize."

"Show me the $50 problem."

After I showed the problem, Alexei replied:

"There are much easier ways to earn $50!"

Would you like to see this problem? You can find it (Problem E5.8) in chapter E5 of *Further Explorations* in this book. This problem is still open, and in the 1994 version of this book I have increased the prize for its first solution to $100!

A. Soifer, *The Colorado Mathematical Olympiad and Further Explorations:* *From the Mountains of Colorado to the Peaks of Mathematics,* DOI 10.1007/978-0-387-75472-7_9, © Alexander Soifer, 2011

As for the $20 problems, in early 1990, $40 was awarded to Vladimir Baranovsky, a high school student from the Siberian city of Omsk, who in 1990 entered Moscow State University[1]. The remaining $20 was awarded to Royce Peng from Palos Verdes High School in California, who in 1990 became a student at Harvard University[2]. These two prize winners met while representing the Soviet Union and the United States respectively at the International Mathematical Olympiad in Beijing, China in July, 1990 (as I predicted in my book [S2] completed half a year earlier!).

One more summer student of mine, Boris Dubrov from Minsk, Belarus, who in 1990 entered Minsk University[3], apparently solved all three $20 problems. His first letter, containing three correct answers, made it through to me. The second letter, containing solutions, never came. It was probably intercepted by overzealous lovers of the spying game of the cold war style.

During that August 1989 program, I spent long hours with Boris, a wonderful human being and a talented young mathematician. He told me about a visit to Moscow by the celebrated American mathematician (and now my friend) Ronald L. Graham. During his interview with the Russian mathematics magazine *Kvant*, Professor Graham mentioned a beautiful problem that dealt with 2-colored positive integers. Boris generalized the problem to n-coloring, strengthened the result, and proved it all! He gave me this problem for the Colorado Mathematical Olympiad.

The problem was the celebrated Schur Theorem of 1916, rediscovered by Boris, with his own proof that was much better than Issai Schur's original proof, but which was already known as well (it appeared in [GRS], and with the theorem's history and proof in [S11]).

Chances of receiving a solution of such a problem during the Olympiad were very slim. Yet, the symbolism of a Soviet kid sending an astonishingly beautiful problem (and solution!) to his American peers was so great that I decided to include this problem as an additional "Problem 6." We have never before – or since – used the

---

[1] Vladimir is now an assistant professor of mathematics at the University of California, Irvine [added in Springer's 2011 edition].

[2] In c. 2005, when I last exchanged information with Royce, he was a Ph.D. student at the University of Southern California [added in Springer's 2011 edition].

[3] Boris is now senior researcher at the Belorussian State University [added in Springer's 2011 edition].

6th problem in the Olympiad. You can find the stronger version of it and a couple of other great problems in chapter E10 of *Further Explorations* in this book.

My other August 1989 student, Vladimir Baranovski, contributed an easy problem 7.1. Problem 7.4 came from the famous problem solver, Professor Joel Spencer of the Courant Institute of Mathematics, New York.

Problems 7.3 and 7.5 were created especially for this Olympiad by me. In fact, Problem 7.5 was born while I was showing Gary Miller, a fabulous high school teacher, and a key judge of the Olympiad, how one can create a problem.

In an unprecedented show of interest in mathematics, 970 junior and senior high school students from all across Colorado came to the University for the Seventh Olympiad. Participants came from Fort Collins and Pueblo, Boulder and Castle Rock, Littleton and San Luis, Woodland Park and Cheraw, La Junta and Florissant, Kiowa and Longmont, Divide and Fountain, Calhan and Englewood, Manitou Springs and Ramah, Cañon City and Rush, Sedalia and Yoder, San Acacio and Blanca, Parker and Franktown, Fort Garland and Alamosa, Aurora and Elbert, and, of course, Colorado Springs.

It was a wonderful day for Matt Kahle, a junior from a class of Judy Williamson at Air Academy High School. He won first prize which included the gold medal, a $1,000 scholarship, a Hewlett-Packard graphing calculator, my new book *How Does One Cut a Triangle?,* and University and City memorabilia. He was also awarded the special prize for creativity.

Second prizes were awarded to Brian Becker, a junior from Coronado High School; Chris Sprague, a senior from Rampart High School; and Dan Hedges, a sophomore from Mitchell High School. Each of them received the silver medal, a $300 scholarship, a top scientific Hewlett-Packard calculator, the book *How Does One Cut a Triangle?,* and University and City memorabilia.

Third prizes went to Adam Feder of Fort Collins High School, Chris Hall of Cheyenne Mountain High School, Greg Geihsler of Coronado High School, Scott Burles of Manitou Springs High School, Tina Tackett of Liberty High School, and Shawn Ouderkirk of Air Academy High School. Each of them received a Hewlett-Packard calculator, my book *Mathematics as Problem Solving*, and UCCS and City memorabilia.

We also awarded 10 fourth prizes, 35 first honorable mentions, and 120 second honorable mentions.

In addition to our regular speakers at the award presentation, we were honored by a speech by the Colorado Senate Majority Leader Jeffrey M. Wells.

At the award ceremony we premiered our documentary film *Colorado Mathematical Olympiad*. My lecture to honor the winners that year was entitled *"The Art of Problem Solving."*

Problems 7

The problems were selected and edited by the Problem Committee: Gary Miller, Jean Miller, Alexander Soifer, and Robert Wood.

7.1 (*V. Baranovsky*) Given an arbitrary triangle ABC, points M and N are selected inside the triangle so that $m(\angle CAM) = m(\angle MAN) = m(\angle NAB)$ and $m(\angle ACM) = m(\angle MCN) = m(\angle NCB)$, where $m(\alpha)$ denotes the measure of an angle α(in other words, the angles A and C of the triangle are trisected). Prove that MN is a bisector of the angle ANC.

7.2 Prove that among any twelve two-digit numbers there are two whose difference is composed of two identical digits.

7.3 (*A. Soifer*) Numbers $1, 2, \ldots, 1990$ are written in a certain order: $a_1, a_2, a_3, \ldots, a_m, \ldots, a_{1990}$. Then every number a_m is multiplied by its position m in the sequence, and among all the products ma_m the largest one, A, is chosen. What is the smallest value A can have?

7.4 (*J. H. Spencer*) A subset of integers is called *doublefree* if there is no integer x for which both x and $2x$ are in the subset. Find the largest size (i.e., number of elements) of a *doublefree* subset of:

(A) the first 16 positive integers;
(B) the first 100 positive integers.

7.5 (*A. Soifer*) Prove that no matter how the plane is colored in two colors, it contains:

(A) a triangle with the smallest side 1, angles in the ratio 1:2:3, and all vertices colored in the same color;
(B) a triangle with the smallest side 1, angles in the ratio 1:2:4, and all vertices colored in the same color.

7.6 (*I. Schur*) All positive integers are colored in n colors (each number in one color), where n is a positive integer. Prove that there are numbers a, b, and c of the same color such that $a + b = c$.

Solutions 7

7.1. Let us erase AB and BC. We are left with a triangle ANC, in which MA and MC are bisectors (Figure 7.1).

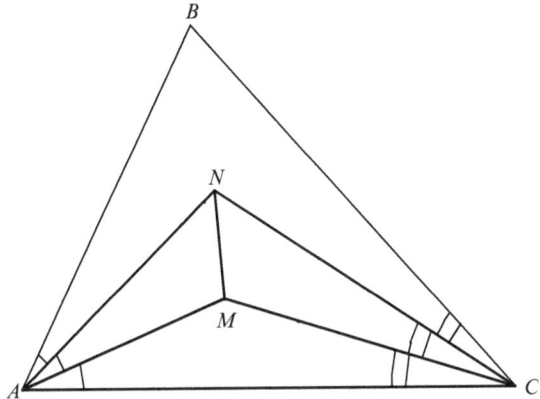

Fig. 7.1

Since the three bisectors of the triangle ANC intersect in one point, M must be that point, and thus MN is a bisector of the angle ANC. ∎

7.2. There are only 11 different remainders upon division by 11: 0, 1, 2, ..., 10; therefore, among any 12 numbers there must be two numbers, call them a and b, that produce the same remainder upon division by 11. But then their difference $a - b$ is divisible by 11. In addition, $a - b$ is a two-digit number. Among the two-digit numbers only those composed of two identical digits are divisible by 11. ∎

7.3(A). All 996 numbers 995, 996, ..., 1990 may not occupy positions 1 through 995 (because then by the Pigeonhole Principle, two numbers would have the same position!). Therefore, at least one of the numbers 995, 996, ..., 1990 has a position no less than 996, and thus:

$$A = \max(ma_m) \geq 995 \times 996.$$

(B) On the other hand, this minimum of A can be attained. Just write the given numbers in the decreasing order 1990, 1989, ..., 996, 995, ..., 1. Then multiply every number a_m by its position m. You can actually compute all 1990 products and verify that for the maximum product A we get

$$A = 995 \times 996 = 996 \times 995.$$

Instead of these bulky computations, you can notice that

$$ma_m = m(1991 - m).$$

It is easy to prove (do) that the product of two factors whose sum is constant, is maximal when the factors are as close to each other in value as possible. This is exactly the situation here: $m + (1991 - m) = 1991$, a constant. Therefore, the maximum of $m(1991 - m)$ is reached when $m = 995$ or 996. ∎

7.4. Let us first look at the following problem. Given a geometric progression P of common ratio 2, i.e., a set of numbers in which every number starting with the second is equal to twice the previous number,

$$P = \{a, 2a, 2^2a, 2^3a, \ldots, 2^{n-1}a\}; (a \neq 0).$$

Find the largest size of a doublefree subset of P.

Obviously, we may not pick two consecutive numbers from P, but we can pick every other number. Thus, the largest size of a doublefree subset of P is $\frac{n}{2}$ if n is even, and $\frac{n+1}{2}$ if n is odd.

Now we are ready to solve problems 7.4(A) and (B).

(A) Partition the set $N_{16} = \{1, 2, \ldots, 16\}$ into 8 disjoint geometric progressions (Figure 7.2).

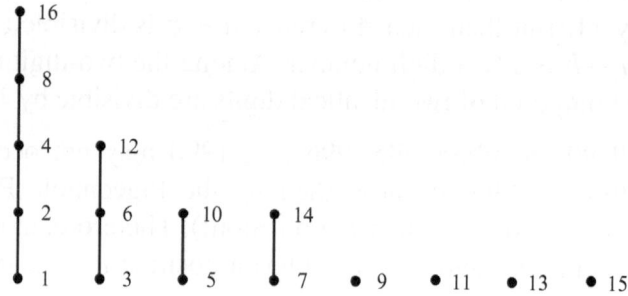

Fig. 7.2

Choices of numbers for a doublefree subset from one geometric progression in no way affect choices from another geometric progression. Thus, we can just choose largest size doublefree subsets from each geometric progression independently, and then combine the choices into a largest size doublefree subset of N_{16}.

The subset {1, 4, 16; 3, 12; 5; 7; 9, 11, 13, 15} is a doublefree subset of the largest size, which is 11.

(B) Similarly to (A), partition $N_{100} = \{1, 2, \ldots, 100\}$ into 50 disjoint geometric progressions (Figure 7.3). Once again we can choose largest size doublefree subsets from each geometric progression independently, and combine the choices into a largest size doublefree subset of N_{100}.

The subset $S = \{1, 4, 16, 64; 3, 12, 48; 5, 20, 80; 7, 28; 9, 36; 11, 44; 13, 52; 15, 60; 17, 68; 19, 76; 21, 84; 23, 92; 25, 100; 27, 29, 31, 33, \ldots, 95, 97, 99\}$ is a doublefree subset of the largest size, which is 67.

Surely S is not the only doublefree subset of N_{100} of size 67. Can you find another doublefree subset of N_{100} of size 67? ■

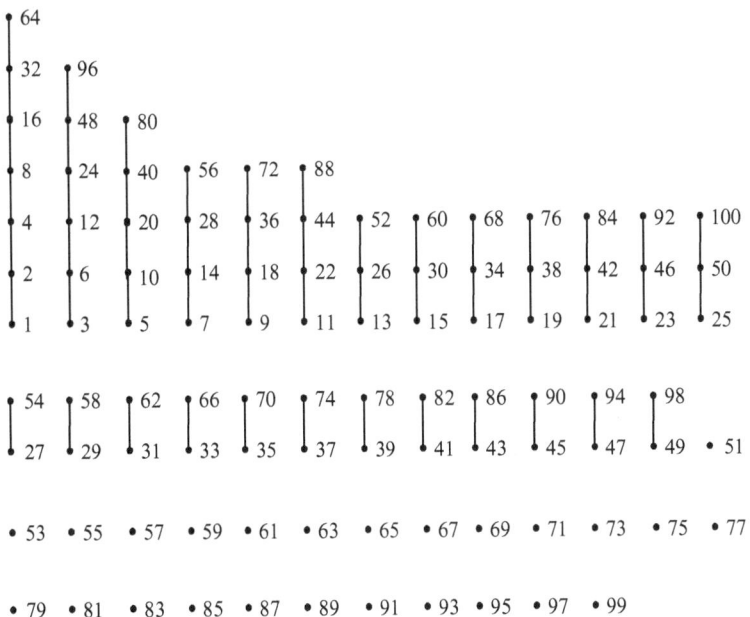

Fig. 7.3

7.5(A). For any positive number d, any 2-colored plane contains two points of the same color distance d apart. This statement was proven in this book for $d = 1$ as Problem 3.2. A proof for an arbitrary d is not much different. We need it here for $d = 2$.

Let P be a 2-colored plane. We can pick in P two points A and B of the same color distance 2 apart and construct a regular hexagon H on AB as on the diameter (see Figure 7.4). Please note, that many triples of vertices of H, for example ABC, form a triangle of precisely the shape described in the problem: a triangle with the smallest side 1 and angles in the ratio 1:2:3.

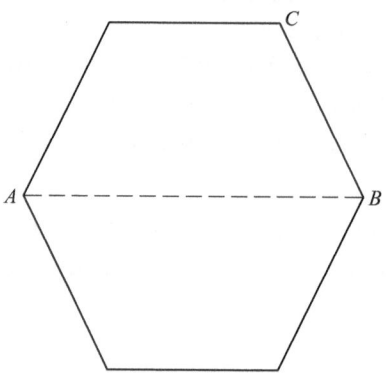

Fig. 7.4

If at least one more vertex, say C, of the hexagon H is of the same color as A and B, then we are done: the triangle ABC *is* a required triangle. If all four other vertices of H have the same color, which is not the color of A and B, then we are done too: *any* three of these four vertices form a required triangle! ■

7.5(B). Assume that such a triangle does not exist in a 2-colored red and blue plane. Toss a regular 7-gon H of side length one on the plane (see Figure 7.5). Observe that many triples of vertices of H, for example ABD, BAF, DFC, DFG, GAC form a triangle of precisely the shape described in the problem: a triangle with the smallest side 1 and angles in the ratio 1:2:4.

Since 7 is an odd number, two adjacent vertices of H must have the same color. Say, A and B are both blue. Then D and F must both be red (for otherwise an all-blue triangle ABD or BAF would provide a contradiction to the initial assumption). Therefore, C and

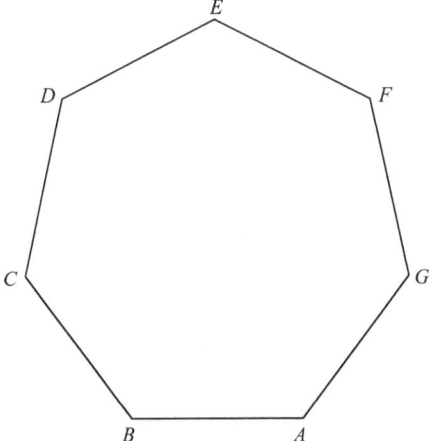

Fig. 7.5

G must both be blue (for otherwise an all-red triangle DFC or DFG would contradict the initial assumption). We have obtained an all-blue triangle GAC in contradiction to the initial assumption. ∎

For more ideas and exciting open problems, see Chapter E6 of *Further Explorations* in this book.

7.6. A *graph* is a finite non-empty set V of vertices some pairs of which (or perhaps none) are said to be *adjacent*. An adjacent pair $e = \{v_1, v_2\}$ of vertices is called an *edge*.

A graph is called *complete* if *every* pair of its vertices is adjacent.

For example, Figure 7.6 presents a complete graph K_5 (the subscript 5 indicates the number of vertices in a complete graph). Sometimes a complete graph K_3 is called a *triangle*.

Please note that the edges connecting pairs of vertices are drawn merely to illustrate that these pairs of vertices are adjacent. The shape of the edges has no meaning at all, and we think of the edges as having no intersection with each other. (You can simply think of a graph as a set of pins with rubber bands connecting some of the pins!)

Let us formulate and prove the following tool:

Tool #3. For any positive integer n there is a number $s(n)$ such that no matter how the edges of the complete graph $K_{s(n)}$ are colored in n colors, the graph contains a triangle with all its three edges colored in the same color.

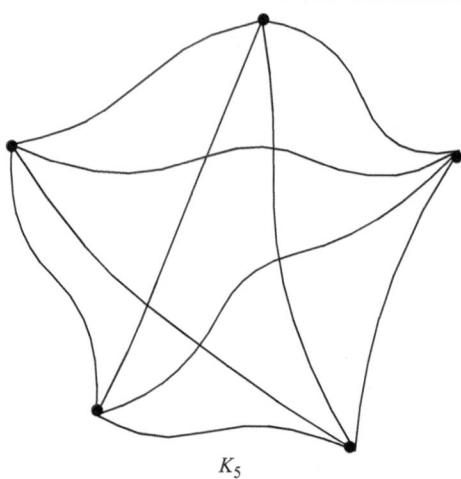

K_5

Fig. 7.6

Proof of Tool #3. We will prove the tool by induction in n.

For $n = 1$ we choose $s(1) = 3$, and the tool is obviously true.

Assume that the tool is true for n, i.e., there is a number $s(n)$ such that no matter how the edges of the complete graph $K_{s(n)}$ are colored in n colors, the graph contains a triangle with all its three edges colored the same color.

We need to prove the tool for $n + 1$. Define

$$s(n + 1) = (n + 1)[s(n) - 1] + 2$$

and look at the complete graph $K_{s(n+1)}$, whose edges are colored in $n + 1$ colors, and pick one of its vertices a. At the vertex a of this graph, $s(n+1) - 1 = (n+1)[s(n) - 1] + 1$ edges originate. Since they are colored in $n + 1$ colors, by the Pigeonhole Principle, we conclude that there is a color, say red, such that at least $s(n)$ edges originating at a are red (Figure 7.7).

Look at the other endpoints $v_1, v_2, \ldots, v_{s(n)}$ of the red edges originating at a. If any two of them, say v_i and v_j, are connected by a red edge, then we are done: all three edges of the triangle av_iv_j are red.

Assume now that no edges connecting the vertices $v_1, v_2, \ldots, v_{s(n)}$ are red. We then have a complete graph $K_{s(n)}$ on these vertices, whose edges are colored in n colors (all $n + 1$ colors but red!). By the inductive assumption, $K_{s(n)}$ contains a triangle with all its edges colored in the same color. Tool # 3 is now proven.

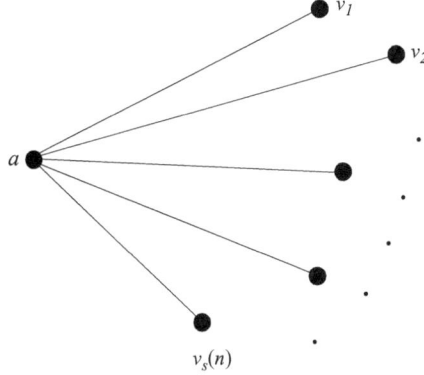

Fig. 7.7

Now we are ready to solve Problem 7.6.

Solution of Problem 7.6. Let all positive integers be colored in n colors $c_1, c_2, \ldots, c_n$. Construct a complete graph $K_{s(n)}$ with the set of numbers $\{1, 2, \ldots, s(n)\}$ serving as its vertices. Now we will color the edges of $K_{s(n)}$ in n colors as follows: let i and j, $(i > j)$, be two vertices of $K_{s(n)}$, we color the edge ij in precisely the color of the number $i - j$ (remember, all positive integers are colored in n colors!). We now have a complete graph $K_{s(n)}$ whose edges are colored in n colors. By Tool #3, $K_{s(n)}$ contains a triangle ijk, $i > j > k$, whose three edges ij, jk, and ik are assigned the same color (Figure 7.8).

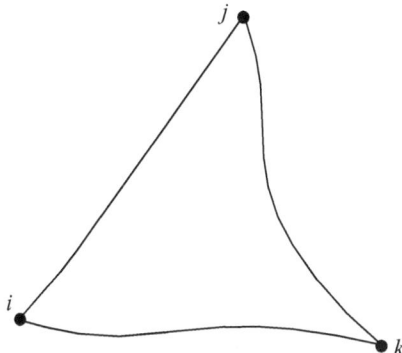

Fig. 7.8

Set $I = i - j$; $J = j - k$; $K = i - k$. Since all three edges of the triangle ijk are colored in the same color, the numbers I, J, and K are assigned the same color in the original coloring of positive integers (this follows from the way we colored the edges of $K_{s(n)}$).

In addition, we have the following equality:

$$I + J = (i - j) + (j - k) = i - k = K.$$

We are done! ■

For a continuation of these ideas, please see Chapter E10 of *Further Explorations* at the end of this book. For much more see [S11], which is completely dedicated to the mathematics of coloring and those who invented it.

Eighth Colorado Mathematical Olympiad

April 26, 1991

Historical Notes 8

The eighth Olympiad attracted the largest number of contestants in the history of the event: 987. As in the previous year, a much greater number of junior high school students participated this time. A good number of junior high school students won awards, especially from the classes of two wonderful teachers: John Putnam and my former student Lossie Ortiz.

For the second consecutive year first prize was awarded to Matt Kahle, now a senior from the class of Judy Williamson at Air Academy High School. Matt received the gold medal, a $1,000 scholarship, a Hewlett-Packard graphing calculator, an autographed copy of a two-week-old book, *Geometric Etudes in Combinatorial Mathematics* by V. Boltyanski and A. Soifer, university and city memorabilia, and a paperweight with a gallium chip from Cray Computer Corporation. Matt is now a student at Colorado State University, Fort Collins. [Added in the 2011 Springer edition: Matt has completed his post-doctoralship at Stanford and in the fall of 2010 will commence his visiting position at the Institute for Advances Study Princeton – see his letter at the end of Part V.]

Dan Hedges, a junior from Mitchell High School, won second prize. Dan received the silver medal, a $500 scholarship, an autographed copy of the 1991 edition of *Geometric Etudes in Combinatorial Mathematics*, a Hewlett-Packard graphing calculator, and memorabilia from the City, the University, and Cray Computer Corporation.

Third prizes were awarded to four young mathematicians: Mike Rauch, a junior from Air Academy High School; Andre Downer, a

A. Soifer, *The Colorado Mathematical Olympiad and Further Explorations: From the Mountains of Colorado to the Peaks of Mathematics*, DOI 10.1007/978-0-387-75472-7_10, © Alexander Soifer, 2011

Contestants at work. *Photograph © by Valari Jack*

junior from Mitchell High School; Lisa Michaels, a senior from the Colorado Springs School; and Chris Dotur, a sophomore from Douglas County High School. Each of them received the bronze medal, a $125 scholarship, a Hewlett-Packard calculator, *Geometric Etudes in Combinatorial Mathematics*, book gift certificates, and memorabilia from the City and the University.

We also awarded 5 fourth prizes, 27 first honorable mentions, and 180 second honorable mentions. The Literary Award was presented to Todd Wakerley, a junior from Coronado High School for his poem "Revenge of the Numbers."

In addition to regular speakers, the award presentation ceremony featured Mayor of Colorado Springs Robert Isaac [today, in 2010, our Airport and Court Building are named in his honor] and engineering manager of Digital Equipment Corporation Karen Pherson. Senator Michael Bird serving again as the Chairman of the Joint Budget Committee of the Colorado General Assembly, also addressed the winners of the Olympiad.

Unable to attend this year's ceremony, Senate Majority Leader Jeffrey M. Wells sent us a letter I read on his behalf:

It is always a special honor to be invited to speak at your Award Presentation Ceremony. This excellent program recognizes and rewards an outstanding group of students and you are to be congratulated for your commitment and support.

I also read a letter addressed to the winners by the Governor of the State of Colorado, Roy Romer:

Congratulations to all of you who participated in, and received honors for, the eighth annual Colorado Mathematical Olympiad. You have been recognized for an important accomplishment and commended for all of your hard work.

Colorado is proud to have students like yourselves who strive hard to excel in math. As you know, good math and problem-solving skills can play an important role in academic and career success. You are proving that your talent can lead us in the future.

On behalf of your friends and neighbors throughout Colorado, I extend congratulations for rising to the challenge and best wishes for continued success.

Keep up the good work.

Right after the Olympiad, on May 20, 1991, I received a very insightful letter from the young Olympian, Jason Oraker, a seventh-grader from Washington Irving Junior High School:

Dear Dr. Soifer,

Thank you very much for putting on the Colorado Mathematical Olympiad. I appreciate that the problems were diverse from the classroom math at my school. They offer a new realm of understanding in mathematics.

I am a seventh grader at Washington Irving Junior High School and am looking forward to competing in the Olympiad in the future.

Thank you also for the very nice calculator and magazine (*Quantum*).

Thanks again,

Jason Oraker

At that time I started writing my *Mathematical Coloring Book*. Accordingly, my award presentation lecture was called *"Problems on a Colored Plane."* I could not imagine then, in 1990, that writing this book would take me 18 years. In 2009, at the age of 18, the book has reached adulthood, and has started living her own life [S11]. One more important thing happened in mid 1991: I founded a research journal, *Geombinatorics*. Let me quote from [S16], where by invitation *Geombinatorics* appeared on the pages of the Russian journal *Mathematical Enlightenment*.

Geombinatorics Is Born

This quarterly journal, started by Alexander Soifer
at the University of Colorado at Colorado Springs,
specializes in geometry and combinatorics, but what
really distinguishes it from the field is attitude!

– Paul Kainen, 1991

Dear Alex,
Finally got around to reading the October Geombinatorics,
which contains more than the usual supply of gems – the bits by
Vizing and Alon were worth more than a year's subscription, if
it's not too crass to put a money value on great mathematics, and
why should it be, art is sold for money all the time ...

– Peter D. Johnson, Dec. 18, 1995

In the spring of 1990 I formulated some problems and conjectures in Euclidean Ramsey Theory, and I wanted to share them with a few colleagues, at least with Paul Erdős, Ron Graham and Branko Grünbaum. How would one do this? Writing several long mathematical letters, but this would take months! Sending Xerox copies, but this would seem to be so impersonal, that I do not know why anyone would read them, let alone think them! :-) And so one day in June of 1990 I called Branko Grünbaum and proposed a solution to this dilemma:

– What do you think about publishing a small lively journal devoted entirely to problem posing essays, work in progress, so that instead of a half a dozen letters to colleagues we would write one essay?

– There are problem sections in many journals, but I do not know a journal completely devoted to open problems. Did you have in mind any limitations?
– How about geometry of combinatorial flavor, *Geombinatorics*?
– A good idea, let us try it.
– But I will need your essays – in every issue.
– OK, I'll try.

Now you know "the rest of the story" of the Conception of *Geombinatorics*, which was born a year later, in June 1991, when its first issue came out. It was tiny, and contained just two essays – by Branko and me. But the volume doubled in the second issue, when Paul Erdős and John Isbell joined the two of us. And then the journal took off. [Now, in October 2010, *Geombinatorics* is on its 20th year.]

So what is *Geombinatorics*? What sets it apart from other journals?

Academic journals remind me of old cemeteries. They publish (bury?), with a great deal of respect, completed research, solved problems. Moreover, the publications appear a couple or more years after the research was completed. By then the results ought to be of not much interest not only for the readers but even for their own authors: years have passed, and an active mathematician would likely be working on something else! *Geombinatorics* is, perhaps, the only publication entirely dedicated to *research in progress*. This is a place to enjoy *live mathematics!*

The style of the journal was reminiscent of Paul Erdős's problem articles. Paul contributed to most issues that came out before his untimely passing in 1996. Numerous essays of Branko Grünbaum, the most frequent author of the journal, have dramatically enhanced the breadth and depth of the journal. Peter Johnson not only contributed his own important essays, but brought to print the celebrated "unpublished" paper, "Colorings of metric spaces," by Benda and Pearls. These three editors as well as our European editors Heiko Harborth of Germany, Janos Pach of Hungary, Jaroslav Nesetril of the Czech Republic, and the American editor Ronald L. Graham made *Geombinatorics* a vital leading journal in problems of discrete and combinatorial geometry.

Geombinatorics has contributed to my lifelong goal of stopping discrimination against young mathematicians based on their age. Our problem posing essays were picked up by graduate students looking for Ph.D. topic (all main results of Paul O'Donnell's Ph.D. thesis were published in *Geombinatorics*). Moreover, we have published research papers by high school and college undergraduate students!

Problems 8

I created problems 8.1, 8.2, 8.3, and 8.5 especially for this Olympiad. Problem 8.4, which originated in the Russian mathematical folklore, was contributed by Alexander Trunov of Dallas, Texas.

The problems were selected and edited by the Problem Committee: Gary Miller, Jean Miller, Lossie Ortiz, Michael Sansing, Alexander Soifer, and Robert Wood.

8.1 (*A. Soifer*) Two policemen are stationed on highway I-25 connecting Colorado Springs and Denver. One officer stationed southbound just before Colorado Springs, gives every passerby $5. The other officer stationed northbound just before Denver, gives every passerby 10 times the money in the passersby's pocket.

You leave one of the cities, Colorado Springs or Denver, go to the other, back to the first, etc., driving 5 legs. Can you end up with exactly $1991? If yes, from what city should you start?

8.2 (*A. Soifer*) Prove that no matter how the vertices of a regular 1991-gon are colored in two colors, there are three vertices of the same color that form an isosceles triangle.

8.3 (*A. Soifer*) Colorado Springs has a bright future. In the XXXth century its subway system will consist of 1991 lines. Every two lines will have exactly one station in common, where passengers can transfer from one line to the other and back. No three lines will have a station in common.

Prove that if on each line we close no more than 10 transfers (but not the stations themselves), then there will be 181 lines such that a passenger can transfer from any one to any other of these 181 lines.

8.4 (*Submitted by A. Trunov*) Prove that any figure of area less than one can be placed on an infinite 1×1 square grid without covering any of the vertices of the grid.

8.5 (*A. Soifer*)

(A) Prove that no matter how the plane is colored in two colors, it contains an $m \times n$ rectangle with all four vertices in points of the same color such that $m = 1$ or 2, and n is a positive integer not greater than 6.

(B) Prove the same statement, with n not greater than 5.

Solutions 8

8.1. If you leave Colorado Springs with $1, then after 5 legs you would end up with $1991. Why? Because

$$11[11(11 \times 1 + 5) + 5] = 1991. \quad \blacksquare$$

8.2. Let P be a 1991-gon whose vertices are colored in two colors, say red and blue. Since 1991 is an odd number, there are two adjacent vertices A and B of the same color, say, red.

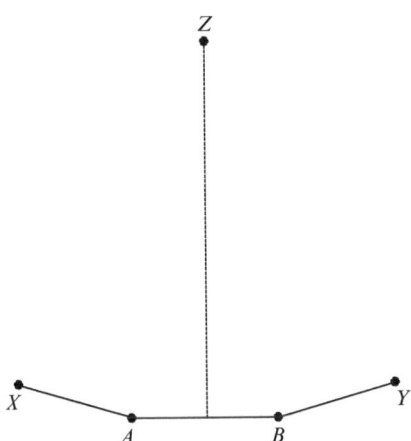

Fig. 8.-3

Let us denote by X the other vertex adjacent to A, and by Y the other vertex adjacent to B (please see Figure 8.-3).

We will denote by Z the vertex of the 1991-gon equidistant from A and B (it exists because 1991 is an odd number).

If at least one of the vertices X and Y is red, then we get an all red isosceles triangle XAB or ABY. Let both vertices X and Y be blue. If Z is blue, then the triangle XZY is an all blue isosceles triangle. If Z is red, then the triangle AZB is an all red isosceles triangle, and we are done. ■

For a remarkable story that took place in 1970 during the Soviet Union National Mathematical Olympiad, and an exceptional problem, see Chapter E8 of *Further Explorations* in this book.

8.3. Assume that no more than 10 transfers are closed on each line. Let n be the maximum number of lines such that a passenger can transfer from any one of these n lines to any other of these n lines. Select such a set of n lines. Then each of the remaining $1991 - n$ lines has a transfer closed to at least one line of the set of n lines. On the other hand, all these closings are achieved by closing at most 10 stations on each of the n selected lines. Therefore,

$$10n \geq 1991 - n.$$

Thus: $n \geq 181$. ■

8.4. Toss the figure F of area less than one on the grid. Choose one 1×1 square S of the grid and translate all 1×1 squares of the grid that contain at least one point of F, to coincide with the square S. The lines of the grid divide F into pieces that were located in various unit squares of the grid. The above translations move all these pieces in the square S (Figure 8.-2).

Nevertheless, since the area of the figure F is less than one, all inside points of the square S cannot be covered by pieces of F. Let a be an inside point of S that is not covered, and x a corner of the square S. As you are well aware, instead of moving the figure F, we can move the grid with respect to F. Moreover, if we translate the grid through the vector $\overrightarrow{xa}$ (so that a will become a vertex of the grid), the figure F will not cover any vertex of the grid in its new position (because if it did, it would imply that the point a is covered by a piece of F in contradiction to the choice of a). ■

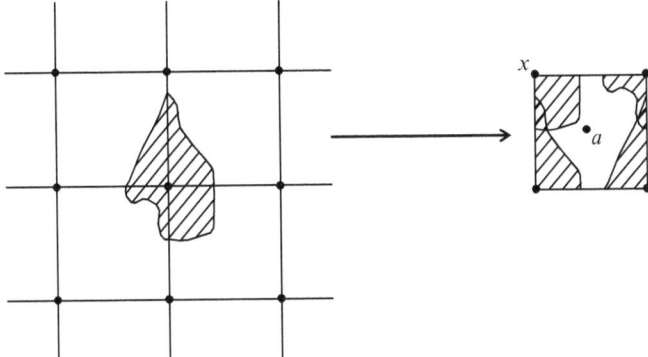

Fig. 8.-2

8.5(A). Toss a 2×6 square grid on a plane colored, say red and blue (please see Figure 8.-1).

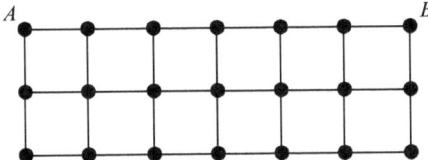

Fig. 8.-1

Out of the 7 vertices in the top row AB of the grid, there are 4 vertices of the same color, say, red. We keep the 4 columns of the grid, that contain these 4 red vertices, and throw away the remaining three columns (please see Figure 8.0).

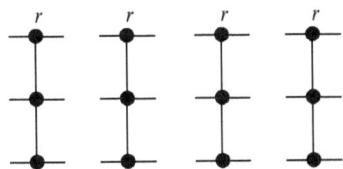

Fig. 8.0

In Figure 8.0 all four vertices of the first row are red. If the second or the third row contains more than one red vertex, then we get an all red rectangle, and the problem is solved.

Assume now that the second and the third rows contain at most one red vertex each. Then by throwing away at most two columns (namely, the columns that contain red vertices in the second or third row), we are guaranteed to be left with at least two columns that have only blue vertices in their second and third rows. Hence, we get an all blue rectangle.

8.5(B). To solve this problem, we need the following tool that is quite useful for other problems as well.

Tool #4. Let the plane be colored in two colors, red and blue, in such a way that there are at least one red and at least one blue points; and let d be a positive real number. Then the plane contains two points of *different* colors distance d apart. (Compare this tool to Problem 3.2).

Proof of Tool #4. Let R, B be red and a blue points in the plane respectively. Draw two circles of radius d with centers at R and B (Figure 8.1). If these circles do not intersect each other, we look at the points R_1 and B_1 of intersection of these circles with the segment $\overline{RB}$.

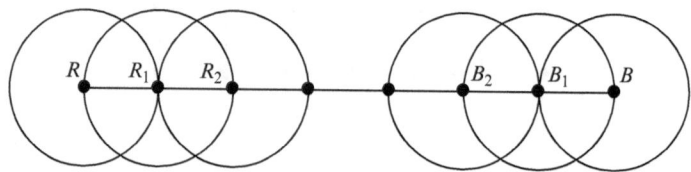

Fig. 8.1

The point R_1 must be red and B_1 must be blue, for otherwise we are done. Now draw circles of radius d with centers at R_1 and B_1. If these circles do not intersect each other, we look at the points R_2 and B_2 of intersection of these circles with the segment $\overline{R_1 B_1}$. The point R_2 must be red and B_2 must be blue, or else we are done. We continue this process until the circles of radius d with centers at R_n and B_n intersect each other at a point C (Figure 8.2).

If C is red, then C and B_n deliver the required pair. If C is blue, then C and R_n are all we are after.

Solution of 8.5(B). If the entire plane is colored in one color (a particular case of 2-coloring!), then we are obviously done. Otherwise we get to use Tool #4: pick in the plane two points A and B of *different*

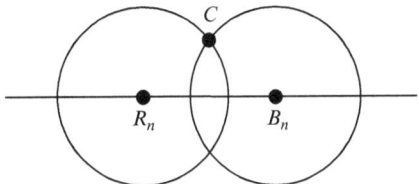

Fig. 8.2

colors distance 6 apart, and construct on them a 2×6 square grid, like the one in Figure 8.1. *Now we can repeat word for word the solution of Problem 8.5(A) in order to prove a stronger statement of 8.5(B)!* Why so? Because when we choose 4 vertices of the *same* color in the row *AB*, we *cannot* choose both vertices *A* and *B* (remember: they are of different colors), and thus the maximal width of the rectangle obtained in the end is reduced from 6 to 5.

Problem 8.5(A), in a different language, was originally created by Professor Cecil Rousseau of Memphis State University, a coach of the American team for the International Mathematical Olympiad. It was then used in the 1975 USA Mathematical Olympiad. In this new setting I was able to reduce *n* from 6 to 5, thus creating problem 8.5(B).■

See Chapter E7 of *Further Explorations* for a related open problem.

Ninth Colorado Mathematical Olympiad
April 24, 1992

Historical Notes 9

This year we featured a truly international set of problems. Problem 9.4 was contributed by the legendary Paul Erdős of the Hungarian Academy of Sciences. Problem 9.3 was inspired by a problem my friend and the leading Abelian group theorist Laszlo Fuchs of Tulane University mentioned in passing as we strolled down his street in New Orleans (Laszlo was born in Hungary). Problem 9.5 was created by Dr. Pak-Hong Cheung of Hong Kong (who created the first inequality in Problem 9.5(A)) and me (the second inequality and Problem 9.5(B)). The two easy Problems 9.1 and 9.2 were suggested by me.

On the day of the Olympiad, Robin Rivers of *The Gazette Telegraph* interviewed Matt Kahle, a former participant and a new judge of the Olympiad, whose name is already familiar to the readers of this book (*Gazette Telegraph*, April 25, 1992 pp. B1 and B3):

It took logic, skill and plenty of imagination to crack the problems.

"It's hard, but it's fun," said Kylan Marsh, a Coronado High School junior.

Matt Kahle, a Pikes Peak Community College student who won the competition in 1990 and 1991 was among the judges. He had competed for five years, watching the contest grow from a few hundred participants to more than 1,000 this year.

"It gives somebody academically talented a chance to shine when there is so much emphasis on athletics in High School," he said.

A. Soifer, *The Colorado Mathematical Olympiad and Further Explorations:*
From the Mountains of Colorado to the Peaks of Mathematics,
DOI 10.1007/978-0-387-75472-7_11, © Alexander Soifer, 2011

The Olympiad reflects an increased emphasis on math and science in US classrooms. President Bush and Gov. Roy Romer have endorsed the goal of America being first in the world in those subjects by 2000.

[Today, in 2010, I can add that President Bush and Governor Romer endorsed ambitious goals, but did nothing to really achieve them. In 2000 – or for that matter in 2010 – the USA has not become first in secondary mathematics education in the world.]

In all previous years, the Olympiad had been proctored almost exclusively by mathematics teachers. This year a new large group, the Palmer High School Alumni Association, headed by Frank and Roberta Wilson, joined this critically important area of the event. They have provided this valuable service to the Olympiad ever since.

Seven hundred students participated in the Ninth Olympiad. First prize was awarded to Dan Hedges, a senior from a class of George Daniels at Mitchell High School. Dan received the gold medal, a $1,000 scholarship, a Hewlett-Packard graphing calculator, an autographed set of my three books, and University and City memorabilia. He also won the special prize for creativity. Upon graduation, Dan entered Southern Methodist University in Dallas, Texas.

Chip Summer, a senior from Cheyenne Mountain High School, won second prize. Chip received the silver medal, a $500 scholarship, an autographed set of my three books, a Hewlett-Packard calculator, and memorabilia from the City and the University.

Third prizes were awarded to four young mathematicians: Scott Mayer, a freshman from Fort Collins High School; Angel Kocovski, a senior from Doherty High School; Lawrence Smith, a sophomore, and Taylor Mohoney, a senior, both from Palmer High School. Each of them received the bronze medal, a $100 scholarship, a Casio or Texas Instruments scientific calculator, *Geometric Etudes in Combinatorial Mathematics* by V. Boltyanski and A. Soifer, a book gift certificate and memorabilia from the City and the University. Scott Mayer also received the United States Space Foundation creativity award, which included a space pen and a one year membership in the Foundation.

We also awarded 4 fourth prizes, 58 first honorable mentions, and 85 second honorable mentions.

Literary Awards were presented to Jacob Lewis for his *Star Wars*-like prose, and to Thomas Katona. Here is the award winning work of the latter.

ODE TO THE MATH OLYMPICS
by Thomas Katona

It was a bright morning up at UCCS
I was ready to succeed, get a 4 and no less.
All the students came in; they were looking their best,
Ready to take this gargantuan test.
The tests were delivered with promptness at nine.
The anticipation was over, I'd finally received mine.
I looked at the problems with awe … my jaw dropped.
With all of my knowledge this test I'd still flop.
The literary competition now this I could win …
That looks brighter, I'll give it a spin.
Next year I'll know to keep up my guard
Cause this test is certainly extremely hard.

In addition to speakers already familiar to readers of this book, the award presentation ceremony was attended by Gay Riley-Pfund, manager of instructional products, Texas Instruments, who came from TI's Dallas Headquarters to greet the winners. New speakers also included Lisa Aré of Colorado Springs City Council and Jack Flannery, Executive Director of the United States Space Foundation. On behalf of Colorado Governor Roy Romer, I read his letter:

It gives me great pleasure to welcome you to the Colorado Mathematical Olympiad awards presentation ceremony.

Colorado takes great pride in the accomplishments of our students. This event is special because it recognizes the academic achievements of young people throughout our state. Through your participation and your achievement, you serve as a role model for fellow students and for other young people in Colorado.

On behalf of the citizens of Colorado, I extend our commendations for your good work and our best wishes for continued success in your studies.

The Engineering Manager of Digital Equipment Corporation, Karen Pherson, was hilarious and profound in her address to the winners. I must have been insane not to include this speech in the first 1994 edition of the Olympiad book. This is a rare perfection. Enjoy:

Once upon a time a far away land was taken over by illiterates. Naturally, mathematicians were held in the greatest of scorn and had trouble getting any form of employment. One such mathematician, who had a Ph.D. from a famous university and had written many learned textbooks, applied for a job as a dishwasher in a local restaurant. The owner asked him for his qualifications and seemed unimpressed even after the mathematician had given him his resume along with an imposing list of publications.

"Tell me a mathematical formula," said the restaurant owner.

"PI R SQUARED," replied the mathematician, referring to the well known formula for the area within a circle.

"I am sorry," said the owner, "You have given me an incorrect formula. Pie are round, cake are square."

Whether you think pies are round or square, Digital Equipment Corporation (or "Digital," as we call it) is especially fond of people like you, people who like mathematics.

In fact, from where we sit we see a whole world of mathematics. Our customers use our computer products very largely to either do math directly or to do other things represented in mathematical ways.

Computers were first developed by mathematicians for their own use and for scientists in universities and government research establishments. They have guided men to the depths of the ocean and supported his journeys to the moon.

Our customers use Digital computers to automate process control in chemical plants, food production plants, breweries, oil refineries and warehouses. Other Digital customers use our products to do finite element matrix analysis of seismic data from oil exploration, to apply mathematical physics of statics and dynamics to the design of cars, ships and bridges; to calculate missile trajectories for national defense; and to calculate proton trajectories for high energy particle research.

And, of course, you are all familiar with the future generations of Digital computers that Captain Picard uses in the Enterprise to navigate his starship through the Galaxy!

We use lots of mathematics to engineer our own products. Our designs and their underlying technologies couldn't be done or even understood without math. For example, we use mathematical models to simulate our circuit and logic designs in order to debug them before we build them.

You can see why we're very interested in you and your math skills. You're going to be our customers, our employees, our stockholders and even our competitors in the years to come. We hope that with a little encouragement from us, added to your talent and enthusiasm for math, you're going to make important contributions to our industry, our community and our world.

So, on behalf of Digital Equipment Corporation, I'd like to congratulate you on your participation and accomplishments in the Ninth Math Olympiad, and wish you well in your future studies.

That year the Colorado Mathematical Olympiad was described in a very important book, the *World Compendium of Mathematics Competitions* [O'H], listing 231 mathematics competitions throughout the world. The Compendium was envisioned and created by Professor Peter J. O'Halloran, President and Founder of the World Federation of National Mathematics Competitions.

Problems 9

9.1 (*A. Soifer*) Is it true that out of any 1992 segments we can always choose three segments that can be used to form a triangle?

9.2 (*A. Soifer*) Every unit square of a 1991 × 1992 checkerboard contains a checker. Alternating turns, two players are allowed to take off the board any number of consecutive checkers from any one column or row. The one who takes off the last checker wins.

Find a strategy allowing one of the players to win regardless of what the other player's moves may be.

9.3 (*A. Soifer, Inspired by L. Fuchs*) Ernő Rubik uses a saw to cut his $3 \times 3 \times 3$ cube into 27 unit cubes. What is the minimal number of cuts he has to make if:

(A) he can only cut one piece at a time.
(B) he can hold a few pieces together in a vise and cut them all at once by one cut.

9.4 (*"Ancient" problem contributed to us by P. Erdős*)

(A) Prove that out of any $n+1$ distinct positive integers not exceeding $2n$, we can always choose three integers x, y, and z (not necessarily distinct) such that

$$x + y = z.$$

Construct an example showing that for n integers this statement is not true.

(B) Prove that out of any $n+2$ distinct positive integers not exceeding $2n$, we can always choose three *distinct* integers x, y, and z such that

$$x + y = z.$$

Construct an example showing that for $n + 1$ integers this statement is not true.

9.5 (*P.-H. Cheung and A. Soifer*) Given n (≥ 2) distinct points in the plane, each pair of these points is joined by a line segment, and its midpoint is colored red. Denote by N the number of red points.

(A) Prove the following inequality:

$$2n - 3 \leq N \leq \frac{1}{2}n(n - 1).$$

(B) For every n find two n-point configurations such that the process described above produces exactly $2n - 3$ red points for one configuration, and $\frac{1}{2}n(n - 1)$ red points for the other.

Solutions 9

9.1. To prove the negative answer, look, for example, at, the set of segments of lengths

$$1, 2, 4, 8, \ldots, 2^{1991}.$$

It is easy to prove (do!) that any three numbers $a < b < c$ from the above list satisfy the inequality

$$a + b < c,$$

which, of course, means that no triangle can be formed. ∎

9.2. Here is one of the winning strategies for the first player. With his first turn, the first player takes off the board all the checkers from the middle row (see Figure 9.1). As a result, we get two identical 995 × 1992 boards, and a player can take checkers off only one of them. Thus whatever the second player takes off one of these boards, the first player takes off the other board. Therefore, the first player will take off the last checker. ∎

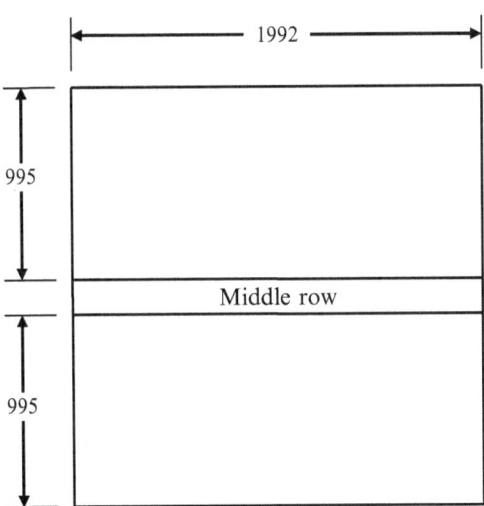

Fig. 9.1

9.3(A). Every time a piece is cut, two pieces are created: the net gain is one piece. Therefore, to go from 1 large cube to 27 small cubes,

26 cuts are required. It surely can be done in 26 cuts. In fact, *any* sequence of 26 consecutive cuts does the job. There is no way to accomplish the job with fewer or with more than 26 cuts!

Note Regarding Problem 9.3(A). For the 2011 Springer edition, I would like to add a very important observation. During the Olympiad, I assumed – as did all the students and judges – that every cut must go all the way through the cube or its part. However, if one allows cuts to be not all the way through (and why not!), surprisingly, we can accomplish the job with a lesser number of cuts! Let us denote by P the plane defined by the two bold lines in Figure 9.2. Make 4 cuts from the front of the cube along the planes of the grid perpendicular to the plane P and 2 units deep. We still have just one piece. Cut all the way through along the plane P. As a result, we get 9 unit cubes and a parallelepiped $2 \times 3 \times 3$. Now make the cut along the plane of the grid all the way through and parallel to P; we get 18 unit cubes and a parallelepiped $1 \times 3 \times 3$. The latter parallelepiped can be cut into unit cubes in just 6 cuts (a 2-dimensional version of the previous set of cuts); I am leaving the pleasure of finding these 6 cuts to you. Thus, we have accomplished the job in the improbable total of just 12 cuts!

■

9.3(B). Observe that in the middle of the $3 \times 3 \times 3$ cube there is one unit cube. While cutting the original cube into unit cubes, we must cut along each of the six faces of this center unit cube, and only one face at a time can be cut. Therefore, the minimum number of cuts is 6. On the other hand, 6 cuts are enough. Just hold the original cube together and make two cuts parallel to each of the three non-parallel faces of the original cube, though the planes of the 3-dimensional grid (one such cut is shown in Figure 9.2). ■

For the 2010 Springer edition, I would like to add an alternative solution of Problem 9.3(B), which I came up with a long time ago.

Second Solution of 9.3(B). After the first cut, we will have a piece P_1 of volume at least 18, no matter what first cut we choose.

The second cut will cut P_1 into at most two pieces, one of which P_2 is of volume at least 9.

The third cut will cut P_2 into at most two pieces, one of which P_3 is of volume at least 5 (the volume must be a whole number!).

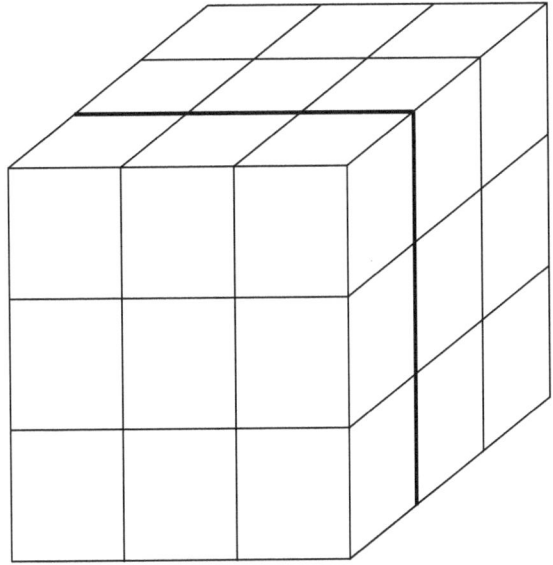

Fig. 9.2

The fourth cut will cut P_3 into at most two pieces, one of which P_4 is of volume at least 3.

The fifth cut will cut P_4 into at most two pieces, one of which P_5 is of volume at least 2. This proves that 5 cuts are not enough, for in the end all pieces must have volume 1! ∎

9.4(A). Arrange the given $n + 1$ integers in an increasing order

$$1 \le a_1 < a_2 < \ldots < a_{n+1} \le 2n$$

and consider the following $2n + 1$ numbers:

$$a_1, a_2, \ldots, a_{n+1}; a_{n+1} - a_1, a_{n+1} - a_2, \ldots, a_{n+1} - a_n.$$

These numbers may not be all distinct, because we just do not have $2n + 1$ distinct integers between 1 and $2n$. Therefore, at least two of them must be equal. The first $n + 1$ numbers $a_1, a_2, \ldots, a_{n+1}$ are given to be distinct. The next n numbers $a_{n+1} - a_1, a_{n+1} - a_2, \ldots, a_{n+1} - a_n$ are distinct as well. Hence, a number from the first group is equal to a number from the second group:

$$a_{n+1} - a_i = a_j$$

for some integers i and j between 1 and n. We are done:

$$a_i + a_j = a_{n+1}. \qquad (*)$$

Observe that n numbers are not enough for the above conclusion. As a counter-example, we can take the set of *large* numbers $\{n + 1, n + 2, \ldots, 2n\}$, or the set of odds $\{1, 3, \ldots, 2n - 1\}$.

9.4(B). This is the same problem as 9.4(A), except we have to insure in the equality (*) that $a_i \neq a_j$. We have an extra $(n + 2)$nd integer for that. Given $n + 2$ integers

$$1 \leq a_1 < a_2 < \ldots < a_{n+2} \leq 2n,$$

if a_{n+2} is even, *and* if among the given numbers there is a_k such that $a_k = \frac{1}{2}a_{n+2}$, then we just *throw away* a_k and then use the reasoning of solution 9.4(A). Otherwise, we can throw away *any* one number and then repeat the reasoning of solution 9.4(A).

As for a counter-example showing that $n + 1$ numbers are not enough, take the set of *large* numbers $\{n, n + 1, \ldots, 2n\}$. ■

9.5(A). I will present two solutions of this problem. The second solution is shown after the solution of 9.5(B).

First Solution. For n distinct points in the plane there are precisely $\frac{1}{2}n(n - 1)$ segments connecting the points. Therefore, the number of midpoints N of these segments satisfies the upper bound of the problem: $N \leq \frac{1}{2}n(n - 1)$.

To prove the lower bound, choose a line L that is not perpendicular to any segment connecting two of the given n points, and project all n points on L. This choice of the line L guarantees that we will get n distinct points on L that are projections of the n given points.

Of course, the midpoint r of the segment connecting two given points a and b would be projected into the midpoint r' of their projection $a'b'$ (Figure 9.3).

Thus, *it suffices to prove the lower bound for the particular case when all n points lie on a line*! We can prove the lower bound $N \geq 2n - 3$ for n distinct points on a line by induction.

For $n = 2$ we get $N_2 = 1 = 2 \times 2 - 3$, i.e., the lower bound holds for $n = 2$.

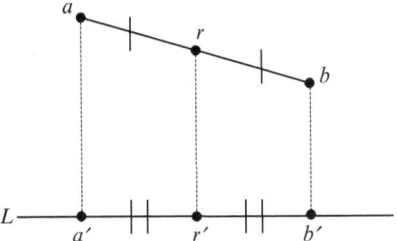

Fig. 9.3

Assume that for any n distinct points on a line, $n \leq 2$, the number N_n of red points satisfies the inequality

$$N_n \geq 2n - 3.$$

Given $n + 1$ distinct points $a_1, a_2, \ldots, a_{n+1}$ on a line (please see Figure 9.4), throw away the last point a_{n+1}.

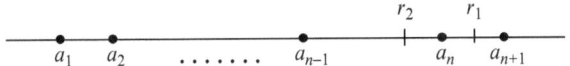

Fig. 9.4

By the inductive assumption, the points $a_1, a_2, \ldots, a_n$ produce $N_n \geq 2n - 3$ red points. Now we can put the point a_{n+1} back. It produces at least two new red points: the midpoints r_1 and r_2 of the segments $a_{n+1}a_n$ and $a_{n+1}a_{n-1}$ respectively. Hence, the number N_{n+1} of red points produced by the $n + 1$ given points satisfies $N_{n+1} \geq 2n - 3 + 2 = 2(n + 1) - 3$. Thus, the desired lower bound holds for $n + 1$.

The lower bound is proven.

9.5(B). You can easily verify (do!) that n evenly spaced points on a line produce exactly $2n - 3$ red points (Figure 9.5).

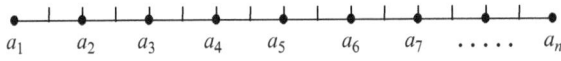

Fig. 9.5

To show that the upper bound can be realized, draw a number line and choose n points with coordinates $2^1, 2^2, \ldots, 2^n$ on it (Figure 9.6).

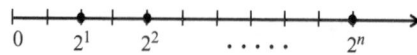

Fig. 9.6

We have to prove that the midpoints of the segments, defined by the pairs of the chosen points, may not coincide. Assume the opposite, i.e., that the midpoints of two segments $[2^a, 2^b]$ and $[2^c, 2^d]$ coincide, where a, b, c, and d are all distinct positive integers, and assume for definiteness that a is the largest of the four exponents. This means precisely that

$$\frac{1}{2}(2^a + 2^b) = \frac{1}{2}(2^c + 2^d),$$

or
$$2^a + 2^b = 2^c + 2^d,$$

hence
$$2^a < 2^c + 2^d.$$

On the other hand,

$$2^a = (2^{a-1} + 2^{a-2} + \cdots + 2^1 + 1) + 1 > 2^c + 2^d.$$

This contradiction proves that no midpoints coincide. Therefore, in this example we have indeed $\frac{1}{2}n(n-1)$ red points.

9.5(A). *Second Proof of the Lower Bound.*

Out of n given points choose two, call them a and b, that are *furthest distance apart*. Segments connecting a to $n-1$ other points, produce $n-1$ red midpoints; plus segments connecting b to the remaining $n-2$ points (b is already connected to a) produce $n-2$ more red midpoints:

$$(n-1) + (n-2) = 2n - 3.$$

All that is left to prove is that none of the first $n-1$ red points coincides with one of the latter $n-2$ red points. Assume they do coincide in the red point r (please see Figure 9.7, in which c and d are two of the given n points).

We get a parallelogram $abcd$.

An easily provable theorem of elementary geometry (try to prove it on your own) states that for any parallelogram,

$$D_1^2 + D_2^2 = K^2 + L^2 + M^2 + N^2,$$

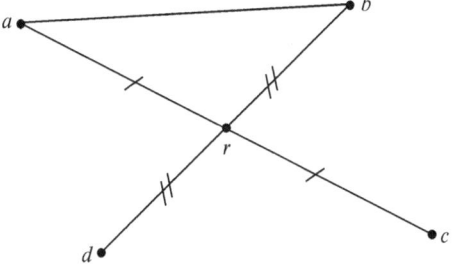

Fig. 9.7

where D_1 and D_2 are lengths of diagonals, and K, L, M, and N are lengths of sides of the parallelogram.

This implies that at least one of the diagonals ac or bd of our parallelogram is longer than its side ab. This contradicts the selection of a and b as two of the given points that are furthest distance apart!

We have proved that $N \geq 2n - 3$.

This striking short solution occurred to me during the Olympiad. Unlike the inductive proof above, this solution explains why the lower bound is what it is. This is not always the case in mathematics. The proof of the Four Color Theorem, for example, does not explain why four colors suffice for coloring any map. ∎

Tenth Colorado Mathematical Olympiad
April 23, 1993

Historical Notes 10

A unique sponsor joined the Olympiad this year: the United States Space Command (USSC) and the North American Aerospace Defense Command (NORAD). General Charles A. Horner awarded certificates to the winners of first through fourth prizes, and invited them (plus Sharon Sherman, the Olympiad Administrator, and I) for a tour of NORAD inside Cheyenne Mountain. NORAD proved to be a no joking matter with the banner "Use of Deadly Force Authorized" above the entrance and an underground labyrinth beyond. It was a unique experience for everyone to see the brain of the defense of North America, which also monitors the space near the Earth. We sat at the huge table in the command room where all critical decisions are made with the NORAD commander on duty, the Canadian Air Force Brigadier-General J. C. M. Robért patiently and humorously answering our many questions. I told him that his exciting job may have some drawbacks. Would he be allowed, for example, to visit France today? "Oh yes," he replied, "but tomorrow at 700 hours (i.e., 7 in the morning) I will have to be back here on the job."

USSC and NORAD continued their sponsorship for one more year, I wish it has lasted longer. However, I thank General Charles A. Horner and the many people of the Space Command and NORAD for making these tours possible for the Olympiad winners!

I wish to thank here the four sponsors that have supported the Olympiad all ten years from its inception. Three are high tech companies Hewlett-Packard, Digital Equipment Corporation, and Texas

A. Soifer, *The Colorado Mathematical Olympiad and Further Explorations:* *From the Mountains of Colorado to the Peaks of Mathematics,* DOI 10.1007/978-0-387-75472-7_12, © Alexander Soifer, 2011

Instruments.[1] The fourth sponsor has supported the Olympiad anonymously all these ten years: John Wicklund is a mathematics teacher. He was a teacher of the first Olympiad winner Russel Shaffer, and many other prize winners of the Olympiad. John has served as a proctor practically every Olympiad during the first decade, and for every year has anonymously contributed $100 in book certificates for the winning contestants. "Why have you been doing it?" I asked John with gratitude. "I love my profession. This is my way to give something back to it," replied John.

In the field of 601 participants, first prize was awarded to senior Jon Hsu from a class of Larry Cattell at Green Mountain High School in Lakewood, Colorado. Jon received the gold medal, a $1,000 scholarship, a Hewlett-Packard graphing calculator, an autographed set of my three books, and University and City memorabilia.

Second prizes were awarded to four young mathematicians: Zach Harris from Fruita/Monument, Kevin Landmark from Coronado High School, Ben Olmstead from Parker Vista Middle School, and Kevin Wiatrowski from Woodland Park High School. Each of them received the silver medal, a $200 scholarship, my book *How Does One Cut a Triangle?*, a Hewlett-Packard graphing calculator, and memorabilia from the City and the University. Ben Olmstead, an *eighth grader*, was also awarded a special prize for creativity.

Third prizes were awarded to three contestants: Mark Friedberg from Palmer High School, Matt Baumgart from Arapahoe High School, and Chris Fry from Air Academy High School. Each of them received the bronze medal, a valuable CASIO overhead graphing scientific calculator, my book, and memorabilia from the City and the University.

We also awarded 6 fourth prizes, 28 first honorable mentions, 27 second honorable mentions, and 78 third honorable mentions. A literary award was presented to the Washington Irving Junior High School bard Dan Wright.

One more time the award presentation ceremony featured the Chairman of the Joint Budget Committee of the Colorado General Assembly, Senator Michael C. Bird, and a letter from Colorado Governor

---

[1] When these lines were published in the first version of this book in 1994, little did I know that soon Hewlett-Packard would unceremoniously stop its support, and Digital Equipment Corporation would end its very existence.

Roy Romer. In addition to regular speakers Dean James A. Null, Dean Margaret A. Bacon, and Past President of the Colorado Council of Teachers of Mathematics John Putnam, the winners were addressed by UCCS Acting Chancellor Merrill J. Lessley, Colorado Springs Councilman John Hazlehurst, and Joy Muhs of CASIO, Inc., who came all the way from Northridge, California.

My award presentation lecture summed up one aspect I especially like about problems of mathematical Olympiads. It was entitled "*It is Easy to See, Especially After Seeing It.*"

I would like to finish my *Historical Notes* of the first decade with a very thoughtful May 5, 1993 letter I received from Colorado State Senate Majority leader Jeffrey M. Wells:

It is always a special honor to be invited to participate in the Math Olympiad. This year I particularly want to personally congratulate you, Dr. Soifer, on the 10th Anniversary of the Math Olympiad. For 10 years, the Math Olympiad has sought out students with great proficiency in mathematics. The Olympiad has provided a unique forum for competition in this field, and a place to showcase individual students with extraordinary skills. The Olympiad has created a great opportunity for our students to learn and excel. It is my hope that the Olympiad will continue with its mission well into the Twenty-first Century.

To the students, I want to congratulate each of you for being selected to participate in the Math Olympiad. Your participation alone demonstrates that you have developed complex mathematical skills and have the ability to use these skills better than most of your fellow students, in fact, just about better than anyone. These abilities give you a tremendous advantage in today's society. I want you all to realize how proud your parents and your teachers are of you. Your accomplishments and ambition will take you far in today's world.

Colorado places a major emphasis on the availability and quality of education in the state. The legislature recently approved a bill to spend 2.5 billion dollars on kindergarten through 12th grade education next year. This investment is important because you are the future of Colorado. Our goal is to give you the best possible education with the best teachers available to prepare you for your future.

Your proficiency in mathematics will open doors for you. Your potential to succeed goes far beyond the fields of math and computer science. Having received my undergraduate degree from Duke University in math, prior to attending Law School I know firsthand that your mathematical abilities and logical thinking will allow you to be a success in any career endeavor you choose. I'm sure however that Dr. Soifer hopes many of you will become math and computer science professors. One of the things I've learned is that being successful takes a certain amount of risk and ambition. Success and security don't mix, but calculated risks will give you the ability to be successful.

You all have experienced this already. You have taken the time to work and prepare for this competition, and have taken the risk of competing. You are all successful for taking this risk.

You'll find that the successful people in life are the ones who realize their potential and develop their skills to utilize it. All of you have the potential that leads to success or you wouldn't be here. Your ability in math is only a window through which you can proceed in life.

Problems 10

After 5 years of naming problems "1", "2", ..., "5", I decided in the Sixth Olympiad to give "real" names to the problems, and put a greater emphasis on formulating problems as "real" stories. The names and the stories inspired the contestants, especially the younger ones. Why did I not continue the "story line" in the Olympiads 7-9? I have no idea. But for the tenth Olympiad, I used names and stories again, and stayed with using names ever since. I enthusiastically recommend to all organizers of Olympiads to take the time and care and dress up your problems in fun stories. Our Olympians give problems more effort and more heart when they enjoy our stories!

For the celebration (the Olympiad has survived for a decade!) we intentionally selected problems that looked very much alike. We hoped the Olympians would discover how different they really were. Problems 10.4 and 10.5 were important in their comparison of infinite and finite phenomena. Problems 10.1, 10.3, 10.4, and 10.5 were created by me (and inspired by the Russian mathematical folklore). Problem 10.2 was a version of a problem offered at the Moscow Mathematical Olympiad in 1959; it appeared in a wonderful Russian book [TT].

10.1 1993 SQUARES (*A. Soifer*). Can a square be cut into 1993 pieces each of which is a square?

10.2 FOUR KNIGHTS (*S. L. Tabachnikov and A. L. Toom*, [TT]). Four knights are placed on a 3×3 chessboard: two white knights in the upper corners, and two black ones in the lower corners. In one step we are allowed to move any knight in accordance with the chess rules to any empty square. (One knight's move is a result of first taking it two squares in a horizontal or vertical direction, and then moving it one square in a direction perpendicular to the first direction.)

Is there a series of steps that ends up with the white knights in diagonally opposite corners, and the black knights in another pair of the opposite corners?

10.3 MILLIONS OF PRINCES (*A. Soifer*). Every unit square of a 1993×1993 chessboard contains a *prince*, a playing piece that can move horizontally or vertically to an adjacent square. Is it possible to have all princes make moves at once so that in the end, as in the beginning, every square of the board contains a prince?

10.4 INFINITE – FINITE I (*A. Soifer*).

(A) Each of the *n* unit squares of a given $1 \times n$ strip (*n* is a positive integer) is colored in one color, red or green. We want the following condition (*) to be satisfied:

> (*) *every square has common sides with an even number of oppositely colored squares.*

Accordingly we put a second $1 \times n$ strip above the first one (so a $2 \times n$ rectangle is formed), and color its squares in such a way that the condition (*) is satisfied for the initial strip. If it so happens that the condition (*) is satisfied for all squares of the second strip, then we stop. Otherwise, we put the third strip on top of the second one and color it in such a way that (*) is satisfied for all squares of the second strip, and so on.

Is there a coloring of the initial strip such that this process of adding and coloring consecutive strips will never stop?

(B) Solve the same problem for an infinite (in both directions) initial strip.

10.5 INFINITE – FINITE II (*A. Soifer*).

(A) Every unit square of an infinite square grid is filled with a positive integer in such a way that no matter where on the grid we pick a 3×3 square, the mean of the four numbers in the corners of the square is equal to the integer in its middle. Find the maximum number of distinct integers the grid can have.

(B) Solve the same problem if 3×3 is replaced by 5×5.

(C) Solve the same problem if 3×3 is replaced by 1993×1993.

Solutions 10

10.1. We will present two solutions of this problem.

First Solution. Four squares can be obtained from one square by cutting it along midlines: we realize a net gain of 3 squares (Figure 10.1).

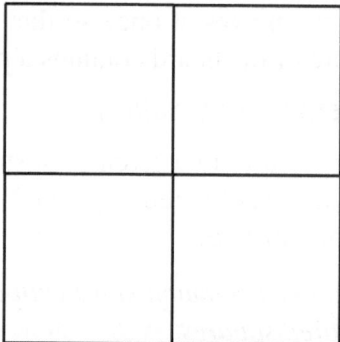

Fig. 10.1

Since $1993 = 3n + 1$ (for $n = 664$), we can cut a square into 1993 squares by consecutively applying the above procedure to squares of partition. ∎

Second Solution. This solution was obtained by several participants of the Olympiad. *It was not known to the organizers!*

Observe:
$$498^2 - 496^2 = 1992.$$

Now the solution is clear: sides of the original square are divided into 498 equal parts, and a square grid is drawn to divide it into 498^2 small squares. Now we combine 496^2 of them (i.e., all little squares except the boundary ones) into one square (please see Figure 10.2). We are done! ■

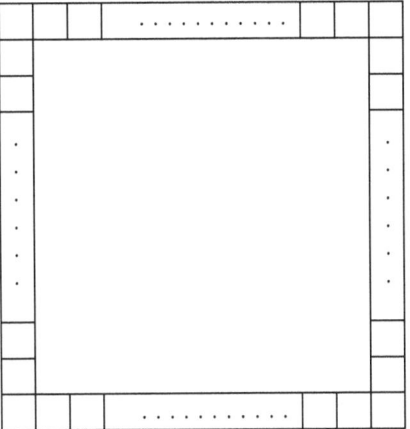

Fig. 10.2

10.2. Number the squares of the 3×3 board and draw the diagram showing how the knights can move (please see Figure 10.3)

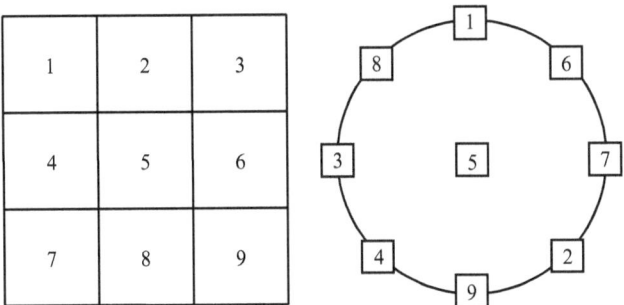

Fig. 10.3 The Moves Diagram

The Moves Diagram shows (to the surprise of many) that knights can move only along the cycle (and never get into square 5). In the initial position, the two white knights are in squares 1 and 3, and the two black knights are on squares 9 and 7, i.e., on the moves diagram the two white knights are followed by the two black knights.

In the desired final position, the white and the black knights would alternate on the moves diagram. To get from the initial position to the desired final position, the order of knights on the moves diagram has to change, which is impossible because the moves diagram is a cycle. ∎

10.3. If we color the 1993×1993 board in a chessboard fashion (Figure 10.4), we will notice that a prince from a black square can move only to a white square, and a prince from a white square can move only to a black square.

The numbers of white and black squares on the 1993×1993 board are *not* equal (because 1993 is an odd number). Therefore, it is impossible to have all princes make moves at once and have every square of the board occupied by a prince. ∎

10.4(A) We will present two solutions to this problem. The second solution is presented after the solution of 10.4(B).

First Solution. We can solve this problem similarly to problem 5.3. Let $k \geq 2$. Observe that the coloring of two consecutive strips $(k-1)$ and k uniquely determines the coloring of the strip $(k+1)$. We can go back as well: the coloring of the strips $(k+1), k$ uniquely determines the coloring of the strip $(k-1)$.

Fig. 10.4

Assume that there is a coloring of the initial strip such that the process of adding and coloring consecutive strips will never stop. Since there are finitely many ways to color a pair of consecutive strips, there is a pair of strips $k, (k + 1)$ that has the same coloring as a pair $(k + d), (k + 1 + d)$ for some positive integer d (please see Figure 10.5).

Fig. 10.5

We can go back and show that the strips $(k - 1)$ and $(k - 1 + d)$ have the same coloring.

Now we have identically colored pairs of consecutive strips $k, (k-1)$ and $(k + d), (k - 1 + d)$. Therefore, the strips $(k - 2)$ and $(k - 2 + d)$ have the same coloring, and so on. We will end up with pairs of strips 2, 1 and $(2+d), (1+d)$ having the same coloring.

The condition (*) of the problem is satisfied for all squares of the strip 1. It will remain satisfied if we add below the strip 1 a new *strip* 0 that is colored in precisely the same way as the strip 1 (Figure 10.6).

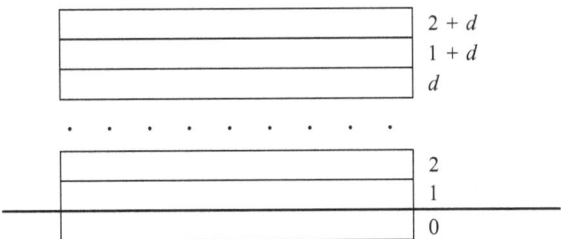

Fig. 10.6

Now we can go back one last time. Since the two pairs of strips 2, 1 and $(2 + d), (1 + d)$ have the same coloring, the strips 0 and d have the same coloring.

But this implies that the strips 0, 1 and $d, (d + 1)$ have the same coloring. This means that, in fact, the condition (*) was satisfied for the strip d before and without putting the strip $(d + 1)$ on top of it! Thus, the process of putting consecutive strips had to stop with the strip d (if not before). This, of course, contradicts our assumption that the process will never stop.

Hence, the process will stop regardless of the coloring of the strip 1. ∎

10.4(B). In the case of an infinite initial strip the answer is the opposite. There is an initial strip such that the process of adding and coloring consecutive strips will never stop. One such example is presented in Figure 10.7.

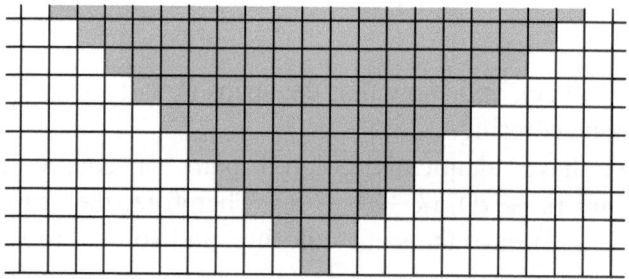

Fig. 10.7

We choose an initial strip with just one red square, and the rest of the squares green. It is easy to prove by induction (do) that this single red square "gives birth" to an infinite red funnel, thus guaranteeing that the process will never stop. ∎

10.4(A) *Second Solution.* No students solved this problem during the Olympiad. However, a number of them, as well as a judge of the Olympiad, Edward Pegg Jr., noticed something not known to me. They noticed, probably after running a few experiments, that *the strip n is precisely the strip 1 in reverse order* (please see an example for $n = 9$ in Figure 10.8).

They also noticed that for an odd n the colored strip number $\frac{n+1}{2}$ (strip 5 in Figure 10.8) is a *palindrome*, i.e., it is symmetric with respect to the central square of the strip.

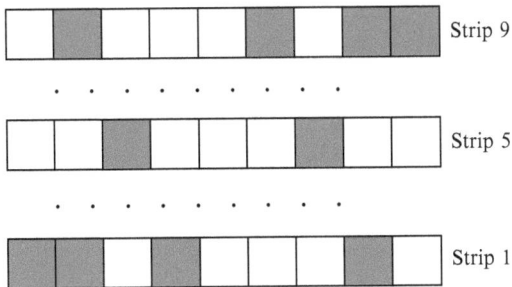

Fig. 10.8

Contestants produced no ideas as to how to prove their conjectures. However, they reinforced Ed Pegg's belief in these conjectures. We were quite busy at the time, reading and grading over 3,000 essays. Nevertheless, Ed had worked on the problem on and off for a couple of hours without producing a proof, when he suddenly declared a brand new conjecture:

All diagonals are palindromes!

Putting strips on top of each other produces a 2-colored square grid of rectangular shape. Ed referred to diagonals of squares in this grid. For some reason this conjecture sounded plausible to me, and I quickly worked out a proof of Ed's conjecture and consequently this second solution of problem 10.4(A).

Tool # 5 (*Conjectured by Edward Pegg, Jr.*). All diagonals are palindromes in the above setting.

I proved it by induction on the length of two consecutive diagonals.

For $m = 1$ the diagonal D_1 consists of just one corner square. Surely, it is a palindrome.

For $m = 2$, the diagonal D_2 consists of two squares a_1 and a_2 adjacent to the square a of the diagonal D_1 (Figure 10.9).

Since a_2 is colored to ensure that a has an even number of adjacent squares of the opposite color, and a has precisely two adjacent squares a_1 and a_2, a_2 had to be colored in the same color as a_1. Thus, the diagonal D_2 is a palindrome too.

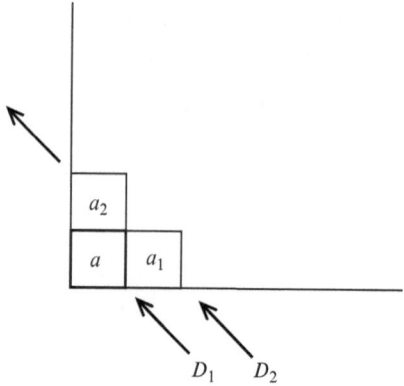

Fig. 10.9

Assume that two consecutive diagonals D_{m-2} and D_{m-1} consisting of $(m-2)$ and $(m-1)$ squares respectively are palindromes ($m \geq 3$). The two diagonals D_{m-2} and D_{m-1} together with the color of the first square c_1 of the diagonal D_m (c_1 is a part of the initial strip) uniquely define the coloring of the diagonal D_m (please see Figure 10.10).

Assume that the squares c_1 and c_m have different colors. Then, since D_{m-2} and D_{m-1} are palindromes (i.e., colors of b_1 and b_{m-1} coincide, colors of a_1 and a_{m-2} coincide, etc.), the squares c_2 and c_{m-1} must have different colors. Similarly, the squares c_3 and c_{m-2} must have different colors. We continue this process until we come to the middle of the diagonal D_m. Since the middle is one or two squares, depending upon the parity of m, we have to consider two cases.

If m is even, then the process described above would produce two adjacent squares $c_{\frac{1}{2}m}$ and $c_{\frac{1}{2}m+1}$ of the diagonal D_m that have different colors (please see Figure 10.11).

Since the colors of the squares $a_{\frac{1}{2}m-1}$ and $a_{\frac{1}{2}m}$ are the same (D_{m-2} is a palindrome), and the colors of $c_{\frac{1}{2}m}$ and $c_{\frac{1}{2}m+1}$ are different, the condition (*) of the problem is *not* satisfied for the square $b_{\frac{1}{2}m}$. We have reached a contradiction.

If m is odd, then the process described above would produce one central square $c_{\frac{1}{2}(m+1)}$ of the diagonal D_m that is surrounded by the two squares $c_{\frac{1}{2}(m-1)}$ and $c_{\frac{1}{2}(m+3)}$ of different colors (please see Figure 10.12). Since the colors of the squares $a_{\frac{1}{2}(m-3)}$ and $a_{\frac{1}{2}(m+1)}$

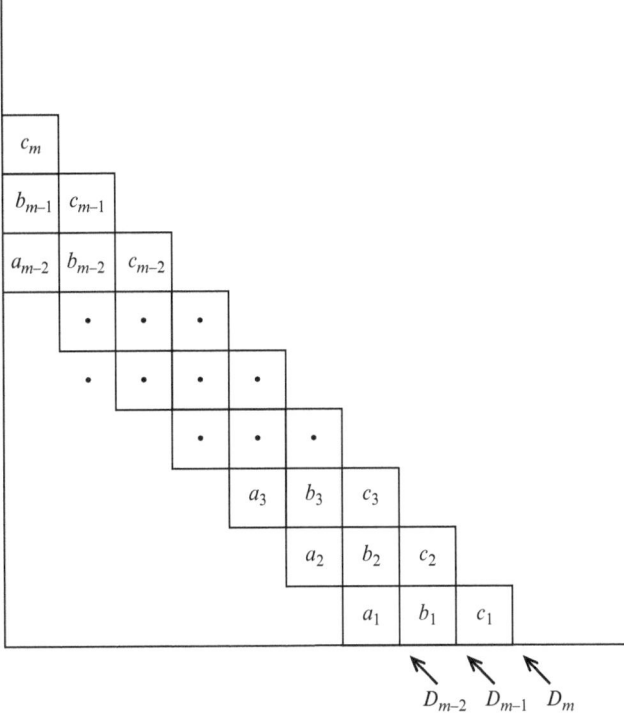

Fig. 10.10

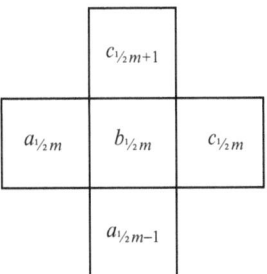

Fig. 10.11

are the same (D_{m-2} is a palindrome), we come to the conclusion that the condition (*) of the problem cannot be satisfied for both $b_{\frac{1}{2}(m-1)}$ and $b_{\frac{1}{2}(m+1)}$, a contradiction.

Both cases ended up in contradictions. Therefore, we have proved that the squares c_1 and c_m have the same color. We can similarly prove

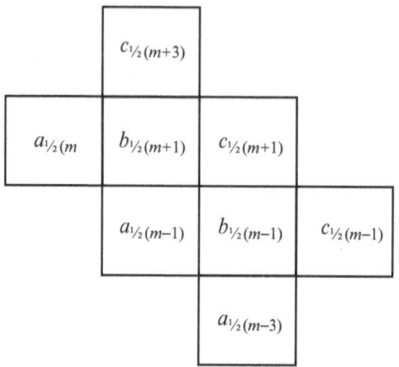

Fig. 10.12

that the squares c_2 and c_{m-1} have the same colors, etc. Thus, D_m is a palindrome. Since D_{m-1} and D_m are palindromes, we can similarly prove that D_{m+1} is a palindrome.

The induction is completed. All diagonals are palindromes. Now I can share with you my second solution.

10.4(A) *Second Solution.*

We need a notation.

Given a column or a row of colored squares with the end squares A and B. Then the symbol $|\overrightarrow{AB}|$ will denote the sequence of colors in this column or row that starts with the color of A and ends with the color of B.

We can now prove that *if the process of putting and coloring consecutive strips does not stop earlier, strip n will be equal to strip 1 in reverse.*

Indeed, assume that we have reached the strip n. Let the corner squares of our $n \times n$ 2-colored grid be A, B, C, and D (Figure 10.13).

Since due to Tool #5, all the diagonals of the grid are palindromes, we get the following equalities:

$$|\overrightarrow{AB}| = |\overrightarrow{AD}| = |\overrightarrow{CD}|,$$

i.e., the strip n is indeed equal to the strip 1 in reverse.

Similarly to the last argument of the first solution of 10.4(A), this means that the condition (*) is satisfied for all squares of the strip n. Thus, the process of putting and coloring consecutive strips stops here, with the strip n (if it has not stopped before!). ∎

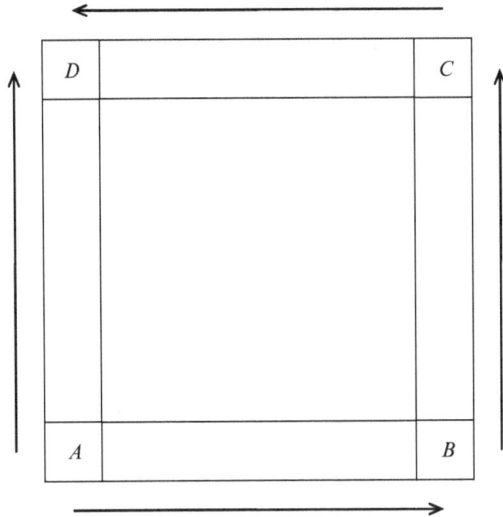

Fig. 10.13

For further related problems, please, see Chapter E9 of *Further Explorations* in this book.

10.5(A) Observe that if we color the grid black and white in a chessboard fashion (Figure 10.3), then the numbers in black squares will provide no restrictions on the numbers in white squares, and vice versa. We have on the grid *two independent systems of numbers.*

Surely the conditions of the problem will be satisfied if we put the same positive integer in all white squares, and another positive integer in all black squares. Thus, we can have 2 distinct integers in the grid.

For convenience, we will call two integers in the grid *neighbors* if they lie in squares of the same color that share a corner.

Let us now prove that the grid may not contain more than two distinct positive integers. Indeed, if it does, then there must be two neighbors a_1 and a_2 that are not equal to each other, say $a_1 > a_2$. Since a_2 is the average of its four neighbors, one of which, namely a_1, is greater than a_2, a_2 must have another neighbor a_3, such that $a_2 > a_3$. We can continue this process indefinitely, thus creating an infinite decreasing sequence of *positive* integers

$$a_1 > a_2 > a_3 > \ldots > a_n > \ldots,$$

which is absurd. ■

10.5(B) This time we color the grid in 8 colors. Please see Figure 10.14, in which every square is labeled by its color, and one square of each of the 8 colors appears inside a figure $\dot{f}$ with bold boundaries. By appropriate translations of $\dot{f}$ we can tile the plane, which determines this nice periodic 8-coloring of the plane.

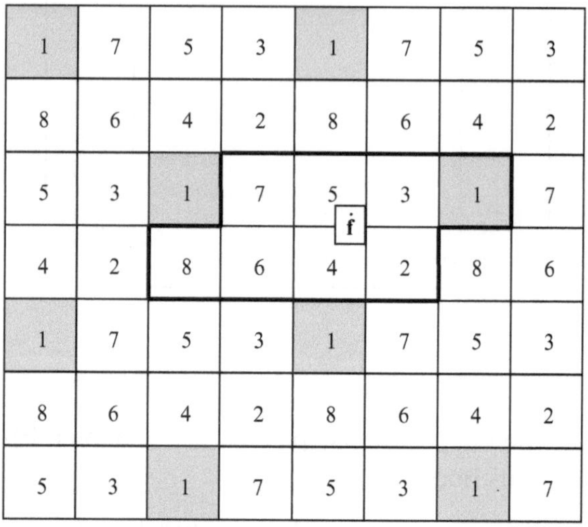

1	7	5	3	1	7	5	3
8	6	4	2	8	6	4	2
5	3	1	7	5	3	1	7
4	2	8	6	4	2	8	6
1	7	5	3	1	7	5	3
8	6	4	2	8	6	4	2
5	3	1	7	5	3	1	7

Fig. 10.14

Observe that the integers in squares of colors 2, 3, ..., 8 provide no restrictions on the integers in squares of color 1. We have on the grid *8 independent systems of numbers*, one for each color.

The conditions of the problem will be satisfied if we put the same positive integer in all squares of color 1, the other positive integer in all squares of color 2, ..., the eighth positive integer in all squares of color 8. We can have 8 distinct integers in the grid.

We will call two integers in the grid *neighbors* if they lie in the squares of the same color closest to each other. Each integer a in the grid has four neighbors, that are in the corners of the 5×5 square whose middle is occupied by a.

By repeating word for word the corresponding part of the solution of problem 10.5(A), we can prove (do!) that the grid may not contain more than 8 distinct positive integers, one for each color.

10.5(C) In this larger setting, the same ideas work. We color the grid in $\frac{1}{2}1992^2$ colors. Unlike the case in the solution of problem 10.5(B), here our figure $\dot{\mathbf{f}}$, which consists of one square of each of the $\frac{1}{2}1992^2$ colors, is pretty large (please, see Figure 10.15).

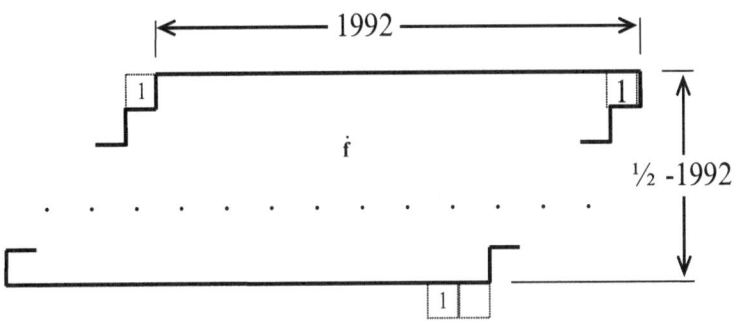

Fig. 10.15

But once again, by appropriate translations of $\dot{\mathbf{f}}$ we can tile the plane, which determines this periodic $\frac{1}{2}1992^2$-coloring of the plane.

By repeating practically word for word the corresponding part of the solution of problem 10.5(A) *or* 10.5(B), we can prove (do) that the maximum number of distinct positive integers in the grid is $\frac{1}{2}1992^2$.

Surely, you noticed the pattern. Verify it by solving problem 10.5(D).

Problem 10.5(D). Solve the same problem if 3×3 is replaced by $(2n - 1) \times (2n - 1)$ for a positive integer n.

Part II
Further Explorations

Science is always wrong. It never solves a problem without creating ten more.

– George Bernard Shaw

I hope that posterity will judge me kindly, not only as to the things which I have explained, but also to those which I have intentionally omitted so as to leave to others the pleasure of discovery.

– René Descartes

Introduction to Part II

When a problem is solved, seldom does a mathematician ask what is next. With his insightful mind the great playwright Bernard Shaw observes that a problem is never solved without giving birth to a number of new problems. My book *How Does One Cut a Triangle?* [S2], [S13] is about this process. It traces the process of going from one problem to another, then another, etc., all the way to the forefront of mathematics, to open problems. In this book we will briefly look at these two trains of mathematical thought (see Chapters E3 and E4 of *Further Explorations*) and several others.

The king is dead, long live the king! The fifty one problems of the Colorado Mathematical Olympiad are solved in the first part of this book. They give birth to a number of more exciting, sometimes open problems that we are going to discuss here in *Further Explorations*. Voltaire believed that "The secret of being a bore is to tell everything." I agree. In this part of the book I will leave out a lot of what is known to me (and everything unknown, of course), "so as to leave to others the pleasure of discovery" (Descartes).

A. Soifer, *The Colorado Mathematical Olympiad and Further Explorations: From the Mountains of Colorado to the Peaks of Mathematics,* DOI 10.1007/978-0-387-75472-7_13, © Alexander Soifer, 2011

E1. Rooks In Space

Inspired by Problem 1.5

Problem 1.5 was another form of the following problem created by Semjon Slobodnik and myself in 1972 and published in 1973, [SS].

E1.1 (*A. Soifer and S. Slobodnik*, 1973). Given 41 distinct two-digit numbers, prove that you can choose five numbers out of them such that any two of the five numbers have distinct units digits and distinct tens digits.

You can certainly represent two-digit numbers as squares of the 10×10 chessboard, a much more visual representation, and thus reduce Problem E1.1 to Problem 1.5. Semjon and I published Problem E1.1 together with the following n-dimensional generalization [SS].

E1.2 (*A. Soifer and S. Slobodnik*, 1973). Given $r \cdot 10^{n-1} + 1$ distinct n-digit numbers, $0 < r < 9$, prove that you can choose $r + 1$ numbers out of them such that any two of the $r + 1$ numbers in any decimal location have distinct digits.

I would not deny you the pleasure of solving this problem on your own by including a solution. Instead, I will tell you what happened next.

In 1986 I taught at the International Summer Institute in the beautiful Hamptons of Long Island, New York. My students included David Hunter, who had just won the Third Colorado Mathematical Olympiad and was on his way to win two more as we all would learn later; a genius, Uri Blass, from Israel, who now holds a doctorate in mathematics, and a group of six very talented students from France. One of

A. Soifer, *The Colorado Mathematical Olympiad and Further Explorations:* *From the Mountains of Colorado to the Peaks of Mathematics*, DOI 10.1007/978-0-387-75472-7_14, © Alexander Soifer, 2011

them was Luc Miller. He solved Problem 1.5 on his own when I offered it in my class, in the same way as had Russel Shaffer, the winner of the First Olympiad. A year later Luc wrote two fascinating letters to me from Paris. Here is a part of the second letter:

> I would like to discuss again this brilliant 41 rooks on 10×10 chessboard problem. You perhaps remember the generalization I had made of it in my last letter. The one you made in language of k-digit numbers [Problem E1.2] is quite different. In fact, I think yours is the good one if you set the problem in terms of numbers, but it does not apply very well to hyper-rooks on hyper-chessboards. The principle of rook movement I retained was: a rook goes straight (as opposed to the bishop who goes diagonally). Now if you associate a number to each rook and you do not want them to attack each other, the rule becomes: two numbers do not differ by only one digit in their decimal representation. Whereas your rule was: two numbers in any decimal location have distinct digits. Note that these two rules are equivalent if there are only two decimal locations. But as shows the illustration on the other page, my generalization is closer to the intuition we have about how a rook should move on a multidimensional chessboard (see the same page for a solution of my generalization).

He went on to formulate and prove the following generalization that was different form Problem E1.2!

E1.3 (*Luc Miller, France, August 11, 1987*). Given $r \cdot 10^{n-1} + 1$ distinct n-digit numbers, $0 < r \leq 9$, prove that you can choose $r \cdot 10^{n-2} + 1$ out of them such that any two of the chosen numbers have at least two decimal locations in which they differ.

Can you solve Luc's problem?

Observe that in both Problems E1.2 and E1.3 we used the decimal system. Surely, the base 10 can be replaced by any positive integer $m \geq 2$. We would get $r \cdot m^{n-1} + 1$ distinct n-digit numbers given with a new boundary for the integer $r : 0 < r \leq m - 1$.

E2. Chromatic Number of the Plane (My Favorite Open Problem)

Inspired by Problem 3.2

This *is* my favorite open problem in all of mathematics, so much so that by the time this new Springer edition is coming out, I have written *The Mathematical Coloring Book* [S11] about this and related problems. In the first 1994 edition of this book I remarked that *The Coloring Book* will appear in 1995, and the following essay is essentially an excerpt from its Chapter II. In fact, *The Mathematical Coloring Book* came out froom Springer much later, on November 4, 2008, the day we elected President Barack Obama. I gave it 18 years and all I had in terms of my knowledge of the arts, sciences, poetry, philosophy, aesthetics, and in terms of all my energy and passion. I hope you will read and enjoy that 640-page magnum opus. I will leave this essay more or less as it originally appeared in 1994, with a new mild editing – it will now serve as an introduction to *The Coloring Book.*

Our good ole Euclidean plane – don't we know all about it? What else can there be after Pythagoras and Steiner, Euclid and Hilbert? In this section we will look at an open problem that exemplifies what is best in mathematics. Anyone can understand this problem; yet, no one has been able to find a solution.

In August 1987 I attended an inspiring talk by Paul Halmos at Chapman College in California. It was entitled *"Some problems you can solve, some you cannot."* This problem was an example of a problem "you cannot solve."

"A fascinating problem...that combines ideas from set theory, combinatorics, measure theory, and distance geometry", say Hallard Croft, Kenneth Falconer, and Richard Guy in their inspiring new book *Unsolved Problems in Geometry* [CFG].

A. Soifer, *The Colorado Mathematical Olympiad and Further Explorations: From the Mountains of Colorado to the Peaks of Mathematics,* DOI 10.1007/978-0-387-75472-7_15, © Alexander Soifer, 2011

"If Problem 8 takes that long to settle (as the famous Four-Color Problem), we should know the answer by the year 2084," write Victor Klee and Stan Wagon in their book *Old and New Unsolved Problems in Plane Geometry and Number theory* [KW].

Are you ready? Here it is.

What is the smallest number of colors with which we can color the plane in such a way that no color contains two points distance 1 apart?

This number is called *the chromatic number of the plane* and is often denoted by χ. A segment in our games here will stand simply for a pair of points. A segment in a colored plane is called *monochromatic* if both its endpoints are colored in the same color.

I do not know who first noticed problem E2.1 – perhaps Adam or Eve? To be a bit more serious, I do not think that ancient Greek geometers, for example, knew this nice fact. They simply did not think of such questions! In terms of the chromatic number of the plane, problem 3.2 can be reformulated as follows.

E2.1 (Adam & Eve?).

$$\chi \geq 3.$$

We can easily prove more:

E2.2 (*E. Nelson, 1950*).

$$\chi \geq 4.$$

i.e., any 3-colored plane contains a monochromatic segment of length 1.

Solution by the Canadian geometers, brothers Leo and William Moser, (1961, [MM]). Toss on the given 3-colored plane what we now call *the Mosers' Spindle* (Figure E2.1). Every edge in the spindle has length 1.

Assume that the seven vertices of the spindle do not contain a monochromatic segment of length 1. Call the colors used to color the plane red, white, and blue. The solution will faithfully follow the children's alphabet song: "A B C D E F G...".

Let the point A be red, then B and C must be one white and one blue, therefore, D is red. Similarly E and F must be one white and one blue, therefore, G is red. We now have a monochromatic segment DG of length 1 in contradiction to our assumption. ■

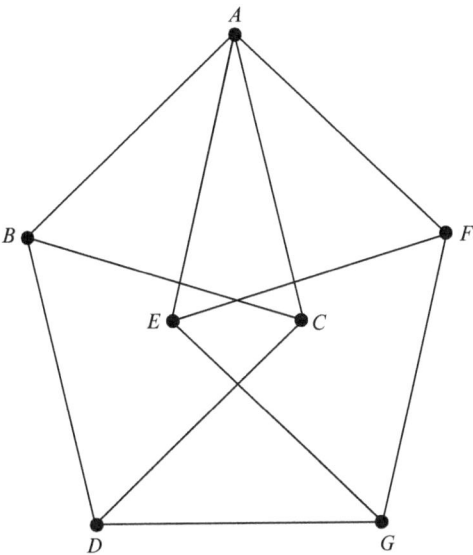

Fig. E2.1 The Mosers Spindle.

When I presented the Mosers' solution to a group of high school students, everyone agreed that it was beautiful and simple. "But how do you come up with a thing like the spindle?" I was asked. As a reply I presented a less elegant but a more naturally found solution. In fact, I would call it a second version of the same solution. Here we touch on a curious aspect of mathematics. In mathematical texts we often see "second solution", "third solution". But which two solutions are called distinct? We do not know. It is not defined, and thus is a judgment call. What may be a distinct solution for one person might be the same solution for another. In a curious coincidence, both versions of this solution were published in the same year, 1961, one in Canada and another in Switzerland. A solution identical to the following one was found by the young American Edward Nelson eleven years earlier. He never published it.

Second Version of the Solution. (*Edward Nelson, 1950; Hugo Hadwiger, 1961,* [H2]). Assume that a 3-colored plane does not contain a monochromatic segment of length 1. Then an equilateral triangle *ABC* of side 1 will have one vertex of each color (Figure E2.2). Let *A* be red. The point *A'* symmetric to *A* with respect to the side *BC* must be red as well.

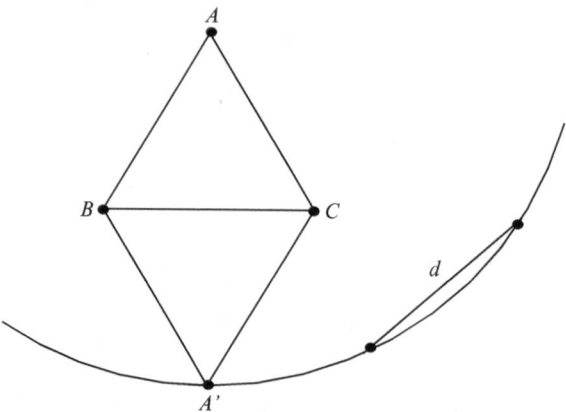

Fig. E2.2

If we rotate our rhombus $ACA'B$ about A through *any* angle, the vertex A' will have to remain red due to the same argument as above. Thus we get a whole red circle of radius AA' (Figure E2.2). Surely it contains a cord d of length 1, in contradiction to our assumption. ∎

Can you find any upper bound for χ?
Think of tiling the plane with regular hexagons!

E2.3 (*J. Isbell*, 1950; *H. Hadwiger* [H1]).

$$\chi \leq 7,$$

i.e., there is a 7-coloring of the plane that does not contain two points of the same color distance 1 apart.

Solution (*J. Isbell*, [H1], [H2]). We can tile the plane by regular hexagons of side 1. Now we color one hexagon in color 1, and its six neighbors in colors $2, 3, \ldots, 7$ (Figure E2.3). The union of these seven hexagons forms a symmetric polygon P of 18 sides. Translates of P (i.e., images of P under translations) tile the plane and determine how we color the plane in 7 colors.

It is easy to prove (please do) that no color has monochromatic segments of any length d, where

$$2 < d < \sqrt{7}.$$

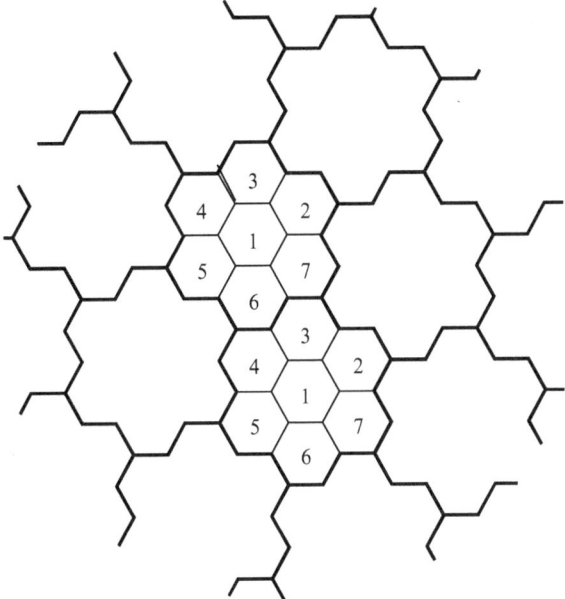

Fig. E2.3

Thus, if we shrink all linear sizes by a factor of, say, 2.1, we will get a 7-coloring of the plane that has no monochromatic segments of length 1. ■

It is amazing that pretty easy Problems E2.2 and E2.3 give us the bounds that still today, in 2010, are best known to mathematics for χ. They were published 49 years ago (in fact, they are 60 years old). Still, all we know is that

$$\chi = 4, \text{ or } 5, \text{ or } 6, \text{ or } 7.$$

A very broad spread! Which do you think it is? Some mathematicians working on this problem think that $\chi = 4$. The legendary Paul Erdős (1913–1996) thought that $\chi > 4$.

Once the late famous American geometer Victor Klee (1925–2007) shared with me a very curious story. In 1980–1981 he lectured in Switzerland. The celebrated Swiss mathematician Van der Waerden was in attendance. When Professor Klee presented the state of this problem, Van der Waerden became very interested. Right there, during the lecture, he started working on the problem. He tried to prove that $\chi = 7$!

What little time I have spent so far working on this problem makes me feel that $\chi = 7$. What a problem! Paul Erdős believed that "God has a transfinite Book, which contains all theorems and their best proofs, and if He is well intentioned toward some, He shows them the Book for a moment." If I ever deserved the honor and had a choice, I would have asked to peek at the page with the chromatic number of the plane. Wouldn't you!

Another amazing thing about this problem is its history, which I researched quite extensively. I have to refer you to *The Mathematical Coloring Book: Mathematics of Coloring and the Colorful Life of Its Creators* [S11] for the complete history of the problem and many fabulous results related to it.

One thing I would like to tell you now. The chromatic number of the plane problem was created in November 1950 by a very young man, the 18-year old Edward Nelson, who has since the 1950s been a Professor of Mathematics at Princeton University (during my 2002–2004 visit to Princeton-Math we became friends). The first person Nelson shared the problem with was the 20-year old John Isbell, who is a Professor of Mathematics at the State University of New York, Buffalo, and remembers their November 1950 meeting well.

E3. Polygons in a Colored Circle, Polyhedra in a Colored Sphere

Inspired by Problem 3.3

With the method used in the solution of Problem 3.3 we can solve other problems as well.

E3.1 A finite number of arcs of a given circle C are painted red. The total length of the red arcs is less than a *third* of the circumference of C. Prove that there is an *equilateral* triangle T inscribed in C with none of its vertices on red.

A third of the circumference is the best possible upper bound for the total length of red arcs:

E3.2 Construct an example of a circle C with the total length of finitely many red arcs equal to one-third of the circumference of C, such that there is no equilateral triangle inscribed in C with none of its vertices on red.

In fact, when you solve problem E3.1, you may notice that the requirement for the triangle T to be *equilateral* is superfluous and can be omitted:

E3.3 A finite number of arcs of a given circle C are painted red. The total length of the red arcs is less than a third of the circumference of C. A triangle T is inscribed in C. Prove that T can be rotated inside C to a position in which none of its vertices lies on red.

A. Soifer, *The Colorado Mathematical Olympiad and Further Explorations: From the Mountains of Colorado to the Peaks of Mathematics*, DOI 10.1007/978-0-387-75472-7_16, © Alexander Soifer, 2011

We can generalize the statement of Problem E3.3 to any inscribed polygon:

E3.4 Let P be an n-gon inscribed in a circle C. The total length of finitely many arcs of C that are colored red is less than $\frac{1}{n}$ of the circumference of C. Prove that the polygon P can be rotated inside C to a position in which none of its vertices lies on red.

We can certainly play these games in space:

E3.5 A map with finitely many regions on a sphere S is colored red and blue. Prove that if the sum of areas of blue regions is less than a fourth of the area of S, then there is a regular tetrahedron inscribed in S with all its vertices in red points; moreover, in inside points of red regions.

More generally, we have the following problem:

E3.6 Let P be a polyhedron with n vertices inscribed in a sphere S. A map with finitely many regions on S is colored red and blue. Prove that if the sum of areas of blue regions is less than $\frac{1}{n}$ of the area of S, then the polyhedron P can be rotated inside the sphere S to such a position that all vertices of P are in red points.

The simple Problem 3.3 gives birth to other explorations as well. One of them leads to the celebrated Borsuk Conjecture; you can find the conjecture and the history of its resolution in [S14]. *The Mathematical Coloring Book* [S11] includes many exciting related problems as well.

E4. How Does One Cut a Triangle?

Inspired by Problem 4.4

The solution of problem 4.4 showed that not every triangle can be cut into two triangles similar to each other, and on the other hand, every triangle can be cut into six triangles similar to each other. In 1970, still an undergraduate student, I posed and solved the following two much more general problems.

E4.1 Find all positive integers n such that every triangle can be cut into n triangles *similar* to each other.

E4.2 Find all positive integers n such that every triangle can be cut into n triangles *congruent* to each other.

It would appear that if anything, the second problem is easier than the first, for similarity is a less restrictive condition than congruency. The opposite, in fact, is true. Problem E4.2 was much harder to solve than Problem E4.1. However, even problem E4.1 was considered by many to be fairly hard. One story comes to mind (I shared it once with my readers in [S2], [S13]). I created both of these problems before April, 1970, when I served as one of the 30 judges of the Soviet Union National Mathematical Olympiad. The judges liked the problem. They selected the critical part of it for the competition of ninth graders:

> Can every triangle be cut into five triangles similar to each other?

The judges meeting to approve the problems was attended by the Chairman of the Organizing Committee, Andrej Nikolaevich Kolmogorov, one of the great mathematicians of the twentieth century.

A. Soifer, *The Colorado Mathematical Olympiad and Further Explorations:* From the Mountains of Colorado to the Peaks of Mathematics, DOI 10.1007/978-0-387-75472-7_17, © Alexander Soifer, 2011

Kolmogorov quietly listened to the presentation of all of the problems and their solutions, and then said:

> "I would only like to replace the problem about five similar triangles."

> "Why Andrej Nikolaevich?" asked the head judge for the ninth grade, Yuri Ionin.

> "It is too difficult. I am not sure I would have solved it," replied Kolmogorov.

The problem was replaced. Of course, I was not thrilled about that. And yet, in a way, I was satisfied. The great Kolmogorov thought that cutting into similar triangles was a difficult problem. "What then would he think about cutting into congruent triangles?" I asked myself.

As promised, I will not deny you the pleasure of solving these exciting problems on your own. I will only list the answers, which may be useful as hints.

Answer for E4.1: All positive integers n, except the first three primes:

$$n \neq 2, 3, 5.$$

Answer for E4.2: All n that are perfect squares, and only them:

$$n = 1^2, 2^2, \ldots, k^2, \ldots$$

For complete solutions and other exciting related problems, I refer you to chapters 1-6 of my book [S2] or better yet its new expanded Springer edition [S13].

E5. Points in Convex Figures

Inspired by Problems 5.4(A), 5.4(B), 5.5(A), and 5.5(B)

During the summer of 1987, I taught at the International Summer Institute in Orange, California. Students came from the U.S., Japan, Israel, Hungary, Switzerland, and France. For their test I decided to create a problem requiring the use of the Pigeonhole Principle in geometry. I came up with Problem 5.4(A). When you solve a problem like that, you ask yourself, can I prove a stronger result, i.e., a result with a smaller n? This train of thought led me to problem 5.4(B), and consequently to problem 5.5(A). The problem became too good to be used for a test. I saved it for the Fifth Colorado Mathematical Olympiad. Problem 5.5(B) shows that the result of problem 5.5(A) is best possible: you cannot reduce n to below 5. Does it mean that we have reached the end of the road? Not at all! Instead of looking at triangles alone, we can include all *convex* figures.

Do you know what *convex* means?

A geometric figure F is called *convex* if for every two points v_1 and v_2 of F, every point of the segment $\overline{v_1 v_2}$ belongs to F.

A triangle and an ellipse are convex; the third figure is not (please see Figure E5.1).

In the spring of 1989, I came up with the following function $S(F)$ defined for any geometric figure F.

Given a figure F, let $S(F)$ denote the minimal positive integer n such that among any n points located inside or on the boundary of F there are always three points that form a triangle of area at most $\frac{|F|}{4}$, where $|F|$ denotes the area of F.

A. Soifer, *The Colorado Mathematical Olympiad and Further Explorations: From the Mountains of Colorado to the Peaks of Mathematics,* DOI 10.1007/978-0-387-75472-7_18, © Alexander Soifer, 2011

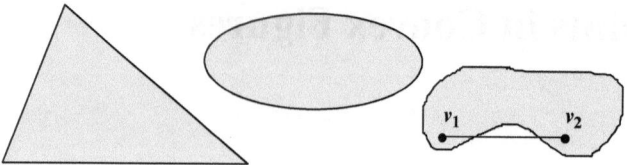

Fig. E5.1

Problems 5.5(A) and 5.5(B) combined can be translated into the language of the function $S(F)$ as follows:

E5.1 For any triangle T,

$$S(T) = 5.$$

Without much difficulty, you can prove the same result for parallelograms:

E5.2 For any parallelogram P,

$$S(P) = 5.$$

Is $S(F)$ equal to 5 for every convex figure? No!

E5.3 For a regular pentagon F,

$$S(F) = 6.$$

Can $S(F)$ assume any other value for a convex figure F? We have an obvious lower bound:

E5.4 For any convex figure F,

$$3 \leq S(F).$$

It is not hard to prove the first upper bound:

E5.5 For any convex figure F,

$$S(F) \leq 9.$$

With more effort you can prove (do) tighter bounds:

E5.6 For any convex figure F,

$$4 \leq S(F) \leq 6.$$

In my proof of the upper bound 6, I used an idea developed *during* the Fifth Colorado Mathematical Olympiad by the winner of one of the two first prizes, Gideon Jaffe, who then received his undergraduate degree from Harvard, Ph.D. in Philosophy from Stanford University, and is now in [2010] an Associate Professor of Philosophy and Law at the University of Southern California.

My intuition was telling me that S(F) cannot be equal to 4 *for any convex figure F.* I posed this conjecture to many mathematicians, and received back two proofs of my conjecture. One came from Dr. Semjon Slobodnik of Moscow, my college classmate and my co-author of Problems 1.5, E1.1 and E1.2 presented in this book. The other proof came from my summer 1987 and 1988 student Royce Peng, who was then at a high school in California, and later entered Harvard University.

It is a difficult problem, but what pleasure you would experience if you solve it!

E5.7 For any convex figure F,

$$S(F) \neq 4.$$

See the proof of this result in [S2] or [S13]. We discovered that the function $S(F)$ is quite amazing: *for any convex figure F,*

$$S(F) = 5 \text{ or } 6.$$

Thus, all convex figures are partitioned into two classes by our function. Would it not be great to know which convex figures belong to which of these two classes! I think so, and accordingly, in the summer of 1989 I offered a fifty dollar prize for the first solution of this problem. The prize is still available today, and in 1999 I raised it to $100!

E5.8 *(One Hundred Dollar Problem).* Find (classify) all convex figures F such that
$$S(F) = 6.$$

You have a chance to win this prize. Try it, but remember: this is an open problem. Nobody has solved it yet. It may be very difficult.

In 1989 I formulated the following conjecture:

E5.9 Let F be a convex figure. $S(F) = 6$ if and only if there is an affine transformation that maps the boundary of F into a fairly narrow frame (width to be determined) made up of two concentric regular pentagons (Figure E5.2).

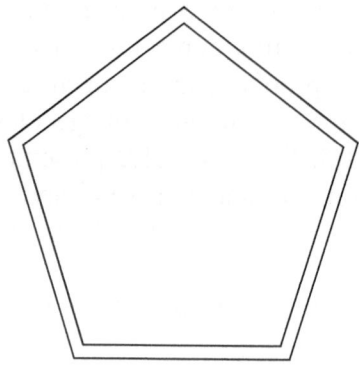

Fig. E5.2

In chapters 7–10 of my books [S2] and [S13], you will find a brief introduction to affine geometry, solutions of the problems above, except the still open Problem E5.8, as well as further problems and generalizations of these ideas. Some of these generalizations are due to Paul Erdős, one of the best problem solvers and the greatest problem poser of all time.

The book [S13] should be of particular interest for you. There I report that recently the young mathematician, a graduate of Columbia University, and now [in 2010] a Ph.D. student at the Courant Institute of Mathematical Sciences, Mitya Karabash disproved my Conjecture E5.9. Problem E5.8 remains open, and Mitya even abstained from proposing his own conjecture!

E6. Triangles in a Colored Plane

Inspired by Problems 7.5(A), 7.5(B)

Problem 7.5(A) and 7.5(B) delivered us examples of triangles that exist *monochromatically* in the plane, no matter how we color it in 2 colors. A triangle in a colored plane is called *monochromatic* if all of its vertices are colored in the same color.

There are many triangles that exist monochromatically in the plane, no matter how we color it in 2 colors. For example, Professor Leslie Shader of the University of Wyoming proved that all right triangles have this property. There is, however, at least one triangle that behaves differently.

This problem and its solution were suggested by the famous Canadian mathematician and problem creator and collector Leo Moser to the world's greatest expositor of mathematics Martin Gardner for his September 1960 *Scientific American* column (the problem appeared in [G1] and its solution in [G2]). It is possible that the late Leo Moser created this problem himself, but I was unable to trace it any further.

E6.1 (*Leo Moser?*). Color the plane in 2 colors in such a way that there is no monochromatic equilateral triangle of side length 1.

Solution ([G2]). Divide the plane into parallel strips, each $\frac{\sqrt{3}}{2}$ wide ($\frac{\sqrt{3}}{2}$ is the altitude of the equilateral triangle of side 1), then color them alternately red and blue (Figure E6.1). Each strip region includes its left border line, and does not include its right border line.

It is easy to verify (do) that our 2-colored plane does not contain a monochromatic equilateral triangle of side 1. ∎

A. Soifer, *The Colorado Mathematical Olympiad and Further Explorations:* 159
From the Mountains of Colorado to the Peaks of Mathematics,
DOI 10.1007/978-0-387-75472-7_19, © Alexander Soifer, 2011

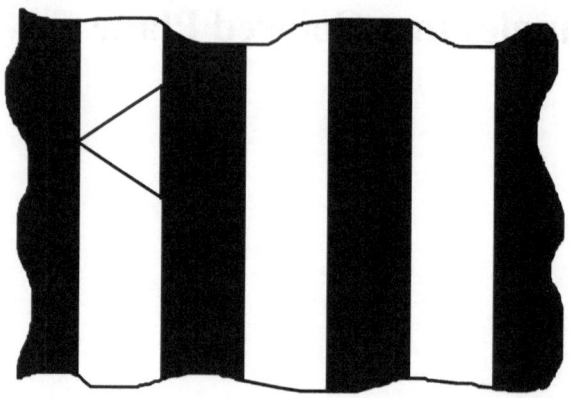

Fig. E6.1

E6.2 Find a coloring of the plane in 2 colors, different from the one in the solution of Problem E6.1, such that the plane does not contain a monochromatic equilateral triangle of side 1.

If you solved Problem E6.2 on your own, you have probably noticed that your and my solution in Figure E6.1 do not differ much from each other. In fact, Paul Erdős et al. thought that the solutions cannot differ much!

Conjecture E6.3. ([EGMRSS], Conjecture 1 of Part III). The only 2-colorings of the plane for which there are no monochromatic equilateral triangles of side 1 are the colorings in alternate strips of width $\frac{\sqrt{3}}{2}$, as in the solution of Problem E6.1, except for some freedom in coloring the boundaries between the strips.

Decades had passed; Ronald L. Graham and Paul Erdős repeated problems and conjectures of Euclidean Ramsey Theory, including E6.3, in their talks and papers (see, for example [E6]), but no proof was found to these easy-looking, hard-to-settle triangular conjectures. However, in March 2006[1] a group of four young Czech mathematicians from Charles University on Malostranské plaza (I visited Jarek Nešetril at this historic place in 1996) Vít Jelínek, Jan Kyncl, Rudolf Stolar, and Tomás Valla [JKSV] disproved this 33-year old conjecture!

---

[1] Of course, this recent result is added in this 2011 version of this book.

Counter Example E6.4. ([JKSV, Theorem 3.19]). Every zebra-like 2-coloring of the plane has a twin 2-coloring that forbids monochromatic unit equilateral triangles.

For definitions of "zebra-like" 2-coloring of the plane and of "twin" coloring, I refer the reader to the original work, which, while not published by *Combinatorica* for years (since submission in March 2006), has been made available by the authors in arXiv. Here I would like to show an example of a zebra-coloring, provided to me by one of the authors, Jan Kynčl (Figure E6.2).

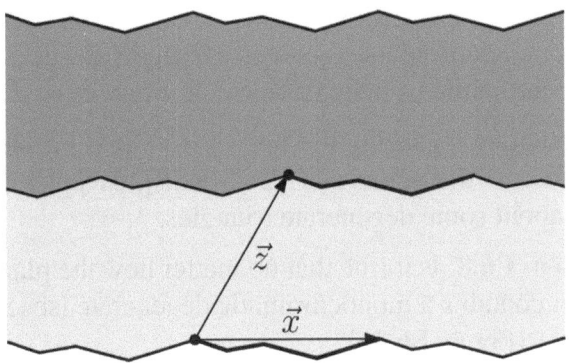

Fig. E6.2

You must be wondering by now which triangles exist monochromatically in the plane no matter how it is colored in 2 colors, and which do not. I would like to know that too. It is still an open problem even though it was first published 37 years ago by a group of six distinguished mathematicians: Paul Erdős, Ronald Graham, Peter Montgomery, Bruce Rothschild, Joel Spencer, and Ernst Straus in their fundamental trilogy [EGMRSS].

Open Problem E6.5. ([EGMRSS]). Find all triangles T such that no matter how the plane is colored in 2 colors, it contains a monochromatic triangle congruent to T.

Paul Erdős et al. made the following conjecture in the same trilogy of papers.

Conjecture E6.6. ([EGMRSS], Conjecture 3 of Part II). For any non-equilateral triangle T, no matter how the plane is colored in 2 colors, it contains a monochromatic triangle congruent to T.

This problem appears surprisingly difficult. Here is what Paul Erdős wrote about it in 1979 [E1]: "Many cases of this startling conjecture have been proved by us (i.e., the authors of [EGMRSS]) and Shader but so far the general case eluded us." In 1985, ([E2]) he put a prize on it: "Is it true that every non-equilateral triangle is 2-Ramsey in the plane (it means exactly the same as Conjecture E6.4)? I offer $250 for a proof or disproof."

Paul Erdős et al. also conjectured that any 2-coloring of the plane does not contain monochromatically at most an equilateral triangle of one size.

Conjecture E6.7. ([EGMRSS]) If the plane colored in 2 colors does not contain a monochromatic equilateral triangle of side d, then it contains a monochromatic triangle of side d' for *any $d' \neq d$*.

My expectations regarding the answer to Open Problem E6.6 are similar to those of the authors of [EGMRSS], except I am not completely sure about some degenerate triangles.

Open Problem E6.8. Is it true that no matter how the plane is colored in 2 colors, it contains a monochromatic degenerate isosceles triangle of small side 1 (Figure E6.3)?

Fig. E6.3

In order to solve Open Problem E6.3 you need tools. Here are two for you. Prove them on your own, then use in your research.

Let T be a triangle. Then $m(T)$ *will stand for the triangle whose sides are twice as long as the corresponding medians of T* (the medians of any triangle are themselves the sides of a triangle – prove this nice elementary fact on your own).

Tool E6.7 ([S4]). For any triangle T, no matter how the plane is colored in 2 colors, it contains a monochromatic triangle congruent to T or to $m(T)$.

Tool E6.8 ([EGMRSS], Theorem 1 of Part III). Let K be a triangle with sides a, b, and c, and let K_a, K_b, and K_c be equilateral triangles with sides a, b, and c respectively. Then the plane colored in 2 colors contains a monochromatic triangle congruent to K if and only if it contains a monochromatic triangle congruent to at least one of the triangles K_a, K_b, K_c.

You received in this section a good deal of open problems and conjectures. I wish you exciting explorations! You will find more results and related open problems in my book [S11].

E7. Rectangles in a Colored Plane

Inspired by Problems 8.5(A), 8.5(B)

The result of problem 8.5(A) was improved in problem 8.5(B). Can we strengthen it further? How much further? It is an open problem. In [S5] I offered a $25 prize for the first solution of this problem. It is still unclaimed in 2010.

Open Problem E7.9. Find the minimum N such that no matter how the plane is colored in 2 colors, it contains a monochromatic $m \times n$ rectangle, such that $m = 1$ or 2, and n is a positive integer not greater than N.

Problem 8.5(B) shows that $N \leq 5$.
I observed the following relationship:

E7.2 Prove that the positive answer to Open Problem E6.8 implies that $N \leq 4$.

A. Soifer, *The Colorado Mathematical Olympiad and Further Explorations:*
From the Mountains of Colorado to the Peaks of Mathematics,
DOI 10.1007/978-0-387-75472-7_20, © Alexander Soifer, 2011

E8. Colored Polygons

Inspired by Problem 8.2

I would like to go back to April of 1970 one more time (Chapter E4 included some reminiscences of that time). The thirty judges of the Fourth Soviet Union National Mathematical Olympiad, of whom I was one, stayed at a fabulous white castle, halfway between the cities of Simferopol and Alushta in sunny Crimea, surrounded by the Black Sea. The problems had been selected and approved in a meeting of all thirty judges and A. N. Kolmogorov. They were being printed. The Olympiad was to take place the next morning, when something shocking occurred.

A mistake was found in the only solution of the problem created by Kolya Vasiliev, a fabulous problem creator, head of the problems section of the journal *Kvant* from its inception in 1970 to his untimely passing in 1998. What were we to do? This question virtually monopolized our lives.

We could just cross this problem out on every one of the several hundred printed problem sheets. In addition, we could add a replacement problem. Both choices were pretty desperate resolutions of the incident. The best, surely, would have been to solve the problem, especially because its statement was quite beautiful, and we had no counter-example to it either.

Even today I can close my eyes and see how each of us, thirty judges, all fine problem solvers, worked on the problem. A few sat at the table as if posing for Rodin's Thinker. Some walked around as if measuring the room. Andrei Suslin, who later solved the famous Serre problem, went out for a hike. Someone was lying on a sofa with his eyes closed. You could hear a fly. The intense thinking seemed to

A. Soifer, *The Colorado Mathematical Olympiad and Further Explorations: From the Mountains of Colorado to the Peaks of Mathematics*, DOI 10.1007/978-0-387-75472-7_21, © Alexander Soifer, 2011

stop the time inside the room. We were unable, however, to stop the time outside. Night fell, and with it our hopes for solving the problem in time.

Suddenly, the silence was interrupted by a victorious outcry: "I got it!" It came from Sasha Lifshits, a student of St. Petersburg University, and former winner of the Soviet Union National and International Mathematical Olympiads (a perfect 42 score at the 1967 IMO in Yugoslavia). His number-theoretic solution used the method of trigonometric sums. However, this was the least of our troubles: the solution was immediately translated into the language of secondary mathematics.

Now we had options. A decision was reached to leave the problem in. The problem and its solution were too beautiful to be thrown away. We knew, though, that the chances of receiving a single solution from several hundred participants were very slim. Indeed, nobody solved it.

It is your chance to try this problem on your own. Heeere's Johnny!

E8.1 (*N. B. Vasiliev*; 1970 *Soviet Union National Olympiad*). Vertices of a regular n-gon are colored in finitely many colors (each vertex in one color) in such a way that for each color all vertices of that color themselves form a regular polygon, which we will call a *monochromatic* polygon. Prove that among the *monochromatic* polygons there are two that are congruent.

I would like to offer a hint (the few, the proud, the marines can ignore it!). The two congruent monochromatic polygons can always be found among the monochromatic polygons with the least number of vertices.

Observe that this problem can be nicely translated into the language of arithmetic progressions that we view to be infinite in both directions.

A sequence infinite in both directions $\ldots, a_{-n}, a_{-n+1}, \ldots, a_0, a_1, a_2, \ldots, a_{n-1}, a_n, \ldots$ is called an *arithmetic progression* if for any integer m, we have the equality

$$a_m = a_{m-1} + k,$$

where k is a fixed real number called *the difference of the arithmetic progression*.

E8.2 Any partition of the set of integers Z into finitely many arithmetic progressions can be obtained *only* in the following way: Z is partitioned into k arithmetic progressions, each of the same difference k (where k is a positive integer greater than 1); then one of these arithmetic progressions is partitioned into finitely many arithmetic progressions of the same difference, then one of these arithmetic progressions (we have progressions of two different differences at this point) is partitioned into finitely many arithmetic progressions of the same difference, etc.

Would you like to read a proof and further history of this problem? You will find them in the first chapter "Merry Go Round" of *The Mathematical Coloring Book* [S11].

E9. Infinite-Finite

Inspired by Problems 5.3, 10.4(A), and 10.4(B)

We looked at two very different solutions of Problem 10.4(A). The first solution was quite powerful. It would stand even if we were to replace the condition (*) by a different one:

(**) Every red square has common sides with an *even* number of green squares; and every green square has common sides with an *odd* number of red squares.

The second solution, on the other hand, showed a deeper insight. We know that the process of adding and coloring consecutive strips in Problem 10.4(A) *will* stop. The question is: *how soon will it stop?* Accordingly, we need to introduce the following notation.

Let $N(n,*)$ be the *smallest* positive integer such that the process of adding and coloring consecutive strips in Problem 10.4(A) will stop with the strip number $N(n,*)$ or sooner, regardless of how the initial $1 \times n$ strip is colored.

The first solution shows merely that

$$N(n,*) \leq 2^{2n}.$$

The second solution proves a much stronger result:

$$N(n,*) \leq n.$$

In fact, an example can be constructed (do) to show that the equality in fact takes place:

$$N(n,*) = n.$$

A. Soifer, *The Colorado Mathematical Olympiad and Further Explorations:*
From the Mountains of Colorado to the Peaks of Mathematics,
DOI 10.1007/978-0-387-75472-7_22, © Alexander Soifer, 2011

Now we can replace the condition (*) in problem 10.4(A) by the condition (**), and ask the following question.

Open Problem E9.1. Let $N(n,**)$ be the *smallest* positive integer such that the process of adding and coloring consecutive strips to satisfy the condition (**) for the previous $1 \times n$ strip, will stop with the strip number $N(n,**)$ or sooner, regardless of how the initial $1 \times n$ strip is colored. Evaluate $N(n,**)$.

We can lift our games up to three dimensions. In fact, we already have three dimensions in Problem 5.3.

E9.2 Solve Problem 5.3 for an infinite (in all four directions) chessboard in place of the $n \times n$ one.

The conditions (*) and (**) can be appropriately translated into the language of three dimensions by replacing the word "squares" by the word "cubes" (in fact, Problem 5.3 used precisely the condition (**) with the word "blocks" in place of "squares"). Accordingly, we can define $N(n^2,*)$ [and $N(n^2,**)$] as the *smallest* positive integer such that the process of putting consecutive layers of n^2 blocks of two colors to satisfy the conditions (*) [respectively condition (**)] for the previous layer, will stop with the layer $N(n^2,*)$ [respectively layer $N(n^2,**)$] or earlier, regardless of how the initial layer is colored.

And the new open problems are:

Open Problem E9.3. Evaluate $N(n^2,*)$.

Open Problem E9.4. Evaluate $N(n^2,**)$.

Those familiar with the n-dimensional Euclidean space E^n, will easily generalize these (and pose related!) problems for E^n.

E10. The Schur Theorem[1]

Inspired by Problem 7.6

As you probably know, in the 1630's, the genius mathematician Pierre de Fermat formulated his so-called Last Theorem in the margin of Diaphantus's *Arithmetica*. He wrote there:

It is impossible to decompose a cube into two cubes, double square [quartic] into two double squares [quartics], and in general, any power greater than square into two powers with the same exponent. I have discovered a truly magnificent proof, but these margins are too small for it.

In other words, equations of the form

$$x^n + y^n = z^n$$

have no solutions in positive integers for values of the exponent n greater than 2.

Fermat's Last Theorem had defied all assaults for over 350 years. When such giants as Leonard Euler failed to prove it, Fermat's Last Theorem became the most famous problem of mathematics. The 40-year old British mathematician, Professor Andrew Wiles of Princeton University in the United States, announced and sketched his proof on June 23, 1993 during his third of a series of three lectures at a conference at Cambridge University. Right at the conference, the Harvard professor Barry Mazur offered Wiles to submit his circa 200-page manuscript to *Inventiones Mathematicae*, where Mazur was an editor.

---

[1] This Exploration has been expanded for this 2011 Springer version of the book.

A. Soifer, *The Colorado Mathematical Olympiad and Further Explorations: From the Mountains of Colorado to the Peaks of Mathematics*, DOI 10.1007/978-0-387-75472-7_23, © Alexander Soifer, 2011

The manuscript was sent to referees, who in August 1993 found a hole in Wiles' argument. Unable to fix the hole himself, Wiles invited his former Ph.D. student, Professor Richard Taylor of Cambridge University, to join in the effort. It took the coauthors a year to find a way to bypass the hole in the original proof. The two papers, by Wiles [W] and by Taylor–Wiles [TW] together occupy completely the 130-page May-1995 issue of *Annals of Mathematics*. (As I am writing these lines, I am looking at a copy of this historic issue that Andrew Wiles inscribed to me on August 2, 2004, as his going away present when I was leaving Princeton after my 2-year visit.) For proving Fermat's Last Theorem, Wiles and his predecessors have created a very important body of mathematics, capable of doing more than settling one great problem.

Previous attempts to prove Fermat's Last Theorem constituted far from a total loss too (keep this in mind as you commence your work in mathematics!). They produced a lot of fabulous mathematics. Issai Schur was one of the many great mathematicians who failed, and in the process created a real mathematical gem in his 1916 paper [SCH].

Issai Schur was born on January 10, 1875 in Mogilev. He was a part of the University of Berlin for most of his academic life, from a student to a full professor ("ordinarius"). It is incredible how popular a professor he was: more than 500 students attended his 1930 course on number theory. Professor Schur made major contributions to algebra, analysis, and number theory. However, as we know, no achievement was high enough for a Jew in the Nazi Germany. Schur was forced into exile, where he died in 1941.[2]

Nobody then asked questions of the kind Issai Schur posed and solved in his 1916 celebrated paper [SCH]. Consequently, nobody appreciated this result much when it was published. The Schur Theorem, which appeared in this book as Problem 7.6, constituted the birth of a new, astonishingly beautiful area of mathematics, now called *Ramsey Theory*.

Boris Dubrov, who proposed Problem 7.6 for inclusion in the Seventh Colorado Mathematical Olympiad, in fact submitted the following result, stronger than Problem 7.6. It is known as well, but what power Boris demonstrated in proving this remarkable result!

---

[2] Read more about Issai Schur and his related results in [S11].

E10.1 (*Stronger Version of the Schur Theorem*). All positive integers are colored in n colors (each number in one color), where n is a positive integer. Prove that there are three *distinct* numbers a, b, and c of the same color such that $a + b = c$.

Proof Let all positive integers be colored in n colors $c_1, c_2, \ldots, c_n$. We add n more colors $c_1', c_2', \ldots, c_n'$ different from the original n colors and construct a complete graph $K_{s(2n)}$ with the set of numbers $\{1, 2, \ldots, s(2n)\}$ serving as its vertices (See the definition of $s(2n)$ in Tool #3 in the solution of Problem 7.6). Now we are going to color the edges of $K_{s(2n)}$ in $2n$ colors.

Let i and j, $(i > j)$, be two vertices of $K_{s(2n)}$, and let c_p be the color in which the number $i - j$ is colored, $1 \le p \le n$ (remember, all positive integers are colored in n colors $c_1, c_2, \ldots, c_n$). Then we color the edge ij in color c_p if the number $\lfloor \frac{i}{i-j} \rfloor$ is even, and in color c_p' if the number $\lfloor \frac{i}{i-j} \rfloor$ is odd (for a real number r, the symbol $\lfloor r \rfloor$, as usual, denotes the largest integer not exceeding r).

We have got a complete graph $K_{s(2n)}$ whose edges are colored in $2n$ colors. By tool #3, $K_{s(2n)}$ contains a triangle ijk, $i > j > k$, whose all three edges ij, jk, and ik are colored in the same color (Figure E10.1).

Set $I = i - j$; $J = j - k$; $K = i - k$. Since all three edges of the triangle ijk are colored in the same color, from the definition of coloring of edges of $K_{s(2n)}$ it follows that in the original coloring of positive integers, the numbers I, J, and K were colored in the same color! In addition we have

$$I + J = (i - j) + (j - k) = i - k = K.$$

We are almost done. We only need to show (our additional promise!) that the numbers I, J, K are all distinct. In fact, it suffices to show that $I \ne J$. Assume the opposite: $I = J$ and c_p is the color in which the number $I = J = i - j = j - k$ is colored. But then the edges ij and jk of the triangle ijk must have been colored in different colors, because

$$\left\lfloor \frac{i}{i-j} \right\rfloor = \left\lfloor 1 + \frac{j}{i-j} \right\rfloor = 1 + \left\lfloor \frac{j}{i-j} \right\rfloor = 1 + \left\lfloor \frac{j}{j-k} \right\rfloor$$

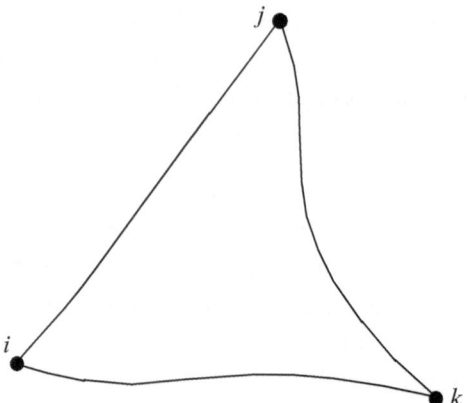

Fig. E10.1

i.e., the numbers $\left\lfloor \frac{i}{i-j} \right\rfloor$ and $\left\lfloor \frac{j}{j-k} \right\rfloor$ have different parity (one is even, and the other is odd). This contradiction to the fact that all three edges of the triangle ijk have the same color proves that $I \neq J$.

Problem E10.1 is solved. How did a high school junior think of such a gorgeous complex solution? Incredible! In 1990 Boris Dubrov entered the University of Minsk. Now, in the year 2010, Boris is senior researcher at the Belarusian State University. ∎

In the Schur Theorem, we proved the existence of a *monochromatic* solution of the linear equation $x + y = z$ (we say that a solution x, y, z is *monochromatic* if all three positive integers x, y, and z are colored in the same color). Can we guarantee monochromatic solutions of non-linear, say quadratic equations? Not so easily, as the following problem shows. It was posed by Paul Erdős and Ronald L. Graham before 1975 (Graham estimates it as "has been open for over 30 years" in his 2005 talk published as [Gr2]).

Open Problem E10.2. (P. Erdős and R. L. Graham, 1975). Does any coloring of the set of positive integers in two colors contain a monochromatic solution x, y, z of the equation $x^2 + y^2 = z^2$?

Graham comments [Gr2]:

There is very little data (in either direction) to know which way to guess.

However, I recall the following story. Following a fabulous July 19-24, 1993, conference at Keszthely on Lake Balaton in Hungary, dedicated to Paul Erdős's 80th birthday, I decided to spend a couple of days in Budapest. In the hotel lobby I ran into the German mathematician Hanno Lefmann from Bielefeld University, Germany. We sipped coffee and talked about coloring problems. As Paul Erdős says, "Mathematicians are machines that transform coffee into theorems." Hanno informed me that he and Arie Bialostocki from the University of Idaho (which is located in *Moscow* (!), Idaho) worked on this Open Problem E10.2. They believed that the answer was negative. In May of 1993 they computer generated a coloring of positive integers from 1 through 60,000 in two colors that forbade monochromatic solutions x, y, z of the equation $x^2 + y^2 = z^2$.

As I mentioned above, Schur's Theorem was the first result in a series of true mathematical gems of Ramsey Theory that includes Baudet-Schur-Van der Waerden's Theorem, Ramsey's Theorem, Rado's Theorem, Hales-Jewett's Theorem, and Graham-Leeb-Rothschild's Theorem. You will find most of these theorems and other magnificent results in *The Mathematical Coloring Book* [S11]. Some of these theorems have been masterfully presented by Ronald Graham in his book [Gr1], which I wholeheartedly recommend.

I would like to finish this essay with one of my favorite results in all of mathematics, which also belongs to Ramsey Theory. It is not widely known even among mathematicians. Its creator was Tibor Gallai (1912-1992), a member of the Hungarian Academy of Sciences. His close friend and co-author Paul Erdős was visiting me when Vera Sós called him from Hungary to communicate the sad news of Gallai's passing.

Gallai discovered a number of fabulous results, some of which were named after other mathematicians: he preferred not to publish even his greatest results. Why? In July 1993 in Keszthely during a dinner that my (then) wife Maya, our baby Isabelle and I shared with George and Esther Szekeres, I was able to ask them about the friend of their youth.

"Gallai was so terribly modest," explained George Szekeres, a famous Australian mathematician and a foreign member of the Hungarian Academy of Sciences. "He did not want to publish because it would show the world that he was clever, and he would be restless because of it."

"But he was very clever indeed," added Esther Klein-Szekeres, a mathematician herself who began her studies in the early 1930's Budapest in a group of young talented Jewish mathematicians: Erdős, Turan, Gallai, Szekeres, and others.

Tibor Gallai discovered this theorem in the late 1930s. He did not publish it either. It appeared in the 1945 paper [R] by Richard Rado (with a credit to Grünwald, Gallai's last name at the time). I hope you will enjoy it, as I do, and try your wit and creativity in proving this beautiful and extremely general, classic result. Readers unfamiliar with m-dimensional Euclidean space can assume $m = 2$: plenty of fun is to be found in the plane!

E10.3 (*Gallai's Theorem,* [R]). Let m, n, k be arbitrary positive integers. If the integer lattice points (i.e., the points with integer coordinates) of the Euclidean space E^m are colored in n colors, and S is any set of k lattice points, then there is a monochromatic set S' (i.e., all points of S' are colored in the same color) of lattice points that is homothetic (i.e., similar and parallel) to S.

In fact, with not too much effort the Gallai Theorem can be strengthened as follows:

E10.4 (*Gallai's Theorem,* [S11]). Let m, n, k be arbitrary positive integers. If the Euclidean space E^m is colored in n colors and S is a k-element subset of E^m, then there is a monochromatic subset S' of E^m that is homothetic to S.

Part III
The Second Decade

The Olympiad: How It Has Continued and What It Has Become

The Colorado Mathematical Olympiad has survived another decade (actually by now it has survived for over a quarter a century). Looking back, I can attest that it has become more sophisticated, more closely linked to research mathematics. At the request of the Executive Director of mathematics at Springer, Ann Kostant, I am adding the second decade of the Olympiad's history and problems in this new Springer edition. However, I cannot just add the Olympiad material without adding its links to and shared essence with "real" mathematics. I therefore am adding 10 new *Further Explorations* to emphasize – and celebrate – the togetherness of the Olympiads and Mathematics. These, now 20, Explorations are the bridges from the Olympiads to mathematical research. Walk across them – and you find yourself in the magical world of mathematics!

A. Soifer, *The Colorado Mathematical Olympiad and Further Explorations:*
From the Mountains of Colorado to the Peaks of Mathematics,
DOI 10.1007/978-0-387-75472-7_24, © Alexander Soifer, 2011

181

The Olympiad: How It Has Continued and What It Has Become

Eleventh Colorado Mathematical Olympiad
April 22, 1994

Historical Notes 11

For the First Time Ever a Freshman Wins

The Olympiad season began on April 18, 1994, when the *Gazette Telegraph* published the article "Olympiad Showcases Math's Beauty, Elegance" by Teresa Owen-Cooper. The journalist offered a case study of the most decorated Olympian David Hunter, and the influence of the Olympiad on his life:

> David Hunter never imagined that he would become what many teen-agers fear most – a math teacher passionate for geometric equations and algebraic formulas.
>
> But that's just what happened to Hunter, 24, who is grateful that a mind-twisting math tournament held every year in Colorado Springs opened his eyes to the mathematical arts.
>
> "It gave me an opportunity to explain things in math," said Hunter, who captured first place in the Colorado Mathematical Olympiad three years in a row before graduating from Palmer High School in 1988. This year's Olympiad will be held Friday.
>
> "It's that creativity I enjoy, fighting the problem and seeing it spark," Hunter said.
>
> Hunter's first success happened when he was just a high school freshman. He was second place – enough to spur his enthusiasm to try again the following year.
>
> "When you compete against your peers and win even second place as freshman, that recognition really encourages you to

A. Soifer, *The Colorado Mathematical Olympiad and Further Explorations:*
From the Mountains of Colorado to the Peaks of Mathematics,
DOI 10.1007/978-0-387-75472-7_25, © Alexander Soifer, 2011

continue," said Hunter, a Princeton University graduate who plans to begin work on master's and doctoral degrees in statistics this fall at the University of Michigan.

The following three years – 1986, '87 and '88 – got even better when he was awarded first place for his solutions to unusual questions that require not only imagination but also speed. Participants have four hours to solve – or try to solve – five problems. Points are awarded for solutions and how the competitors figured the answers.

Hunter said because of the fun he had during the Olympiad, he wanted to pass that on to others. He has been a teacher at Alvirne High School in Hudson, N.H., for two years.

"I believe in public education," said Hunter. "It seems important to me to get good people in teaching. I feel I can contribute and I wanted to give something back."

Alexander Soifer, a mathematics professor at the University of Colorado at Colorado Springs and Olympiad organizer, said this is exactly what he hopes the competition does for today's youth.

"It doesn't matter what profession you choose, if you learn how to think analytically you can go far," said Soifer. "In the last 10 years, society has become very high tech and all this requires a much more sophisticated mind."

Hunter, says Soifer, is a perfect example.

"As a good sportsman, David knew how to plan his time and perform, said Soifer. "He did it very well."

Soifer expects at least 600 participants at this year's competition – started 11 years ago to bring glory to math students by pitting their minds against the maddeningly difficult problems Soifer takes pride in devising.

"American high school education overall is pretty poor compared to France, Russia or a number of other countries, "Soifer said. "An American high school student can finish with one year of physics. Where the school mathematics ends, the Olympiad takes over by offering problems that show the beauty and elegance of math. It really can attract young minds."

The Eleventh annual Colorado Mathematical Olympiad brought together 703 junior high and high school students. Contestants came from all over Colorado: Denver, Parker, Fountain, Cañon City, Woodland Park, Florissant, Divide, Englewood, Littleton, Castle Rock, Manitou Springs, Falcon, Peyton, Fort Collins, Hayden, Longmont, Aurora, Franktown, Cascade, Brush, Merino, Hillrose, Ordway, Olney Springs, Erie, Lafayette, Calhan, Yoder, Rush, Sedalia, Larkspur, Monument, Widefield, and Colorado Springs.

Travis Kopp is receiving his gold medal from Alexander Soifer, while three-time winner David Hunter, who spoke at the Award Ceremony, looks on. Photo by Stewart Wong, *Gazette Telegraph* photographer

In the Olympiad, where the same problems are offered to every participant, from a junior high school student to a senior, it is remarkable when a younger student wins. It is especially admirable when he is just a ninth-grader competing against twelfth-graders. The winner of this year Olympiad, Travis Kopp, was the youngest of all winners in eleven years of the competition. He was a freshman from the class of Mrs. Maculus at Bear Creek High School. Travis solved almost perfectly all problems of the competition, except for the easiest (!) Problem 11.1. At the April 29, 1994 Award Presentation Ceremony Travis received first prize, which included the gold medal, $1,000 scholarship, the set of three autographed books by Alexander Soifer, a $240 Hewlett-Packard graphing calculator, a $25 book gift certificate,

The winner Travis Kopp. Photo by Stewart Wong, *Gazette Telegraph* photographer

and University and City memorabilia. He also received a creativity award, sponsored by Robert Penkhus Jr., the owner of Volvo and Mazda dealerships.

It was a great thrill for me to have at the Awards Ceremony our three-time (1986–1988) Olympiad gold medalist David Hunter address the winners and present together with me the first prize to Travis. David's address was deep and humorous; he started by congratulating especially the winners from the Palmer High School, his alma mater. Our VIP speakers included Colorado Senate Majority Leader Jeffrey M. Welles, UCCS Chancellor Linda Bunnell Shade, and Dean James A. Null.

Second prizes were awarded to three contestants: Lawrence J. Smith, a senior from Palmer High School; Jenifer Wilkes, a sophomore from Liberty High School; and Scott Mayer, a junior from Fort Collins High School. Each of them received the silver medal, a $300 scholarship, a $240 Hewlett-Packard graphing calculator, a $25 book certificate, and University and City memorabilia.

Third prizes were presented to the following eight participants: Mark Friedberg, a junior, Dan Wright, a sophomore, Steve Shell, a freshman, and Jonah Sheridan – all from Palmer High School; Ben Tsai, a freshman from Cherry Creek; Matt Baumgart, a junior from

Arapahoe; Dave Kast and Keith Buck, both seniors from Heritage. Each of them received the bronze medal, CASIO graphing calculator, and University and City memorabilia.

In addition to the above, each of the medalists received the just released first 1994 version of this book, *Colorado Mathematical Olympiad: The First Ten Years and Further Explorations.*

For the second – and alas last – time all our medalists were invited by the General Charles A. Horner, Commander in Chief of the U.S. Space Command and Air Force Space Command, to take a tour of NORAD (North American Airspace Defense Command) inside Cheyenne Mountain.

We also awarded 7 Fourth prizes, 17 first honorable mentions, and 60 second honorable mentions. The Art award was given to John Rondano from Cheyenne Mountain High School. The Literary award was presented to Shawn Peach from Air Academy High School.

Problems 11

Problem 11.4 was created by Professor John Horton Conway of Princeton University and me on March 9, 1994, when we met at a conference in warm and beautiful Boca Raton, Florida. I recall our meeting well: John arrived from snow-covered Princeton in winter boots. He appeared at the conference in shorts and boots and had to quickly buy sandals in the sub-tropical Florida. Problems 11.1, 11.2, and 11.5 were created by me. In fact, I recall the dates of creation for two of the problems: Problem 11.2 was created in April 1989 in Salt Lake City; Problem 11.5 was born on March 30, 1994. I adapted Problem 11.3 from the Russian mathematical folklore.

The problems were selected and edited by the Problem Committee: Gary Miller, Lossie Ortiz, Alexander Soifer, and Donald Sturgill. This year, we assigned weights to problems: Problems 11.1, 11.2, and 11.3 were weighted equally; problems 11.4 and 11.5 carried double that weight.

11.1 HOW DOES ONE CUT A TRIANGLE? (*A. Soifer*). Each side of an equilateral triangle is partitioned into six segments of equal length, and straight lines parallel to the sides are drawn through the points of partition. As a result, we get a triangular grid consisting of thirty-six little triangles congruent to each other. We then cut the original triangle along the lines of the grid into smaller pieces, some of which could be 6-gons or 7-gons.

(A) What is the maximal number of convex 6-gonal pieces we can obtain?

(B) What is the maximal number of convex 7-gonal pieces we can obtain?

11.2 VISIT TO PYTHAGORAS (*A. Soifer*). The circle inscribed in a right triangle T divides its hypotenuse into two segments of lengths a and b. Find the area of T.

11.3 POINTS IN A SQUARE. Ninety-nine points are given inside a 1×1 square. Prove that there is a circle of radius one-ninth that contains at least three of the given points.

11.4 CHROMATIC NUMBER OF A GRID (*J. H. Conway and A. Soifer*). Given a square grid and a positive number d. Is it true that one can always color the vertices of the grid in two colors so that any two vertices of the grid distance d apart are colored in different colors?

11.5 TWO-ROOK GAME FOR TWO PLAYERS (*A. Soifer*). A move in our game consists of placing two rooks on a 1994×1994 chessboard so that no rook placed in previous moves is attacked. Two players take turns. The last player to make a move wins. Find a strategy that allows one player to win regardless of what the moves of the other player may be, if

(A) the two rooks of each move must not attack each other;

(B) the two rooks of each move must attack each other.
   (Two rooks are said to attack each other if they are located in the same row or column of the board.)

Solutions 11

11.1(A). *First Solution.* This solution occurred to me only on June 2, 2002, when I sat down to write a solution for this book. Observe that no matter where a convex hexagon is placed on the given triangle, it will contain completely the smallest regular hexagon the middle of Figure 11.1 or will cover at least one of the striped triangles in the same figure! Hence, there are at most 4 hexagons!

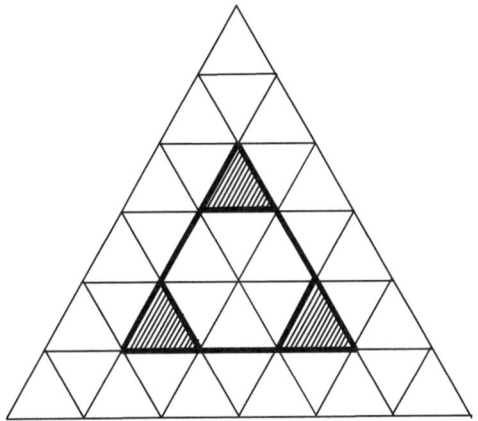

Fig. 11.1

On the other hand, 4 hexagons can be easily obtained (please see Figure 11.2). ■

Fig. 11.2

Second Solution. Look at the strip R of single-triangle "width" adjacent to a side of the given triangle (please see Figure 11.3. At most 6 of its 11 triangles can be included in hexagons. Therefore at least 5 triangles will not be used in hexagons. Counting along all

3 sides, we enumerate at least $5 \cdot 3 - 3 = 12$ unused triangles (we must subtract 3 due to counting corner triangles twice). This means that at most $36 - 12 = 24$ triangles can be used in hexagons, therefore at most $24/6 = 4$ hexagons are possible.

Sufficiency is proven by Figure 11.2. ∎

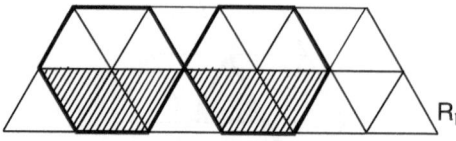

Fig. 11.3

11.1(B). Assume a convex heptagon (i.e., 7-gon) S is cut out of the given triangular grid along the llines of the grid. The sum of all angles of S is $180°(7 - 2) = 900°$. (If you are not familiar with this formula, you can easily prove it by drawing all diagonals of S emanating from one corner and thus partitioning S into 5 triangles.) On the other hand, the maximum angle the triangular grid allows for S is $120°$, and thus the sum of all angles of S is at most $7 \times 120° = 840° < 900°$. This contradiction proves that no convex heptagon can be cut out of the triangular grid along the grid lines, and thus the answer to this problem is 0. ∎

11.2. *First Solution.* Early morning in Salt Lake City, before my talk at the national meeting of the National Council of Teachers of Mathematics, I tried to create a problem that would allow a solution similar to Pythagoras' supposed proof of the Pythagoras Theorem (of course, we know now that the theorem was known 1000 years before Pythagoras, but that is another story). I created such a problem and its three fine solutions. But truly, the first solution was my homage to the great Greek geometer!

Pythagoras partitioned the same square in two different ways, thereby proving his theorem. This can be done in my problem too: we partition the square of side $a + b$.

The first partition (Figure 11.4) produces four rectangles $a \times b$ (each of area ab of course) and a square of side $(b - a)$.

The second partition (Figure 11.5) produces four triangles congruent to the one whose area T we are here to determine, and a square of side $(b - a)$ – the same square that has appeared in the first partition! But this allows us to determine the desired area: $T = ab$. ∎

Second Solution. The rectangle in Figure 11.6 consists of two given triangles put together. Therefore, its area is equal $2T$, where T is the area of the given triangle.

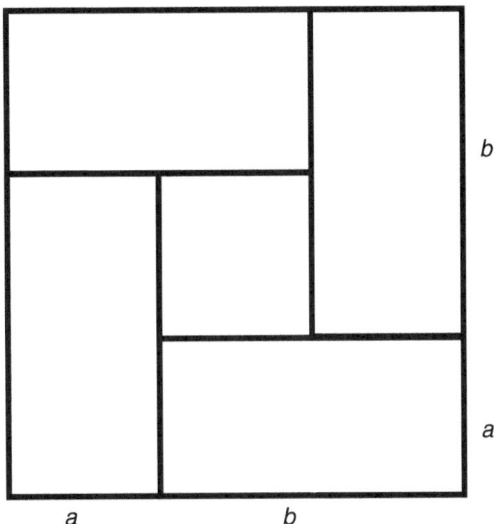

Fig. 11.4

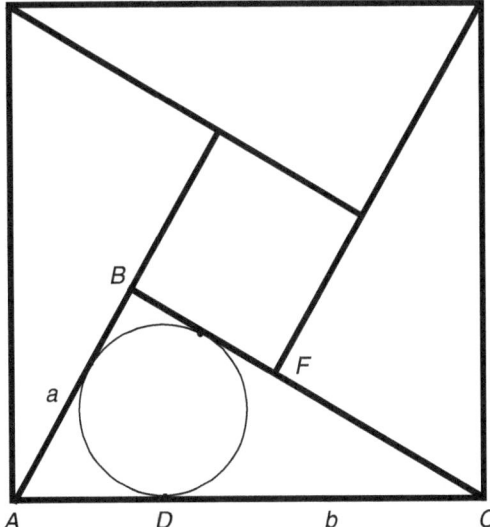

Fig. 11.5

On the other hand, the bold lines partition the large rectangle into a rectangle $a \times b$ (of area ab) plus three more pieces of total area $ar + br + r^2 = (a + b + r)r = T$. The last equality is true for any triangle, not only the given right triangle: the product of the semiperimeter and the radius of the inscribed circle is equal to the area T of the triangle.

Comparing the two partitions of the large rectangle, we get $T = ab.$ ∎

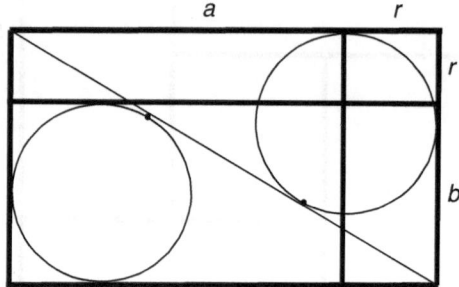

Fig. 11.6

Third Solution. We start here with two formulas for the area T of the given triangle (please see Figure 11.7), one of which appeared in the second solution:

$$(a + b + r)r = T = \frac{1}{2}(a + r)(b + r).$$

Now we can finish up:

$$T = 2T - T = (a + r)(b + r) - (a + b + r)r = ab. ∎$$

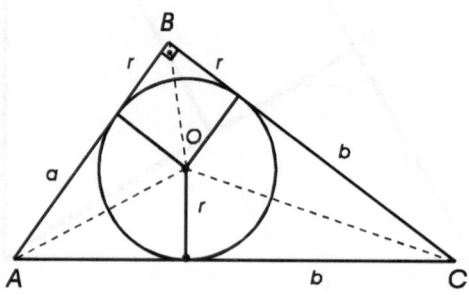

Fig. 11.7

11.3. We need to find a small space that contains at least three points. . . Hmm, sounds like a Pigeonhole Principle problem. We are given 99 points, and 99 can be presented as $99 = 49 \times 2 + 1$. Observe also that 49 is a perfect square. Now a solution is clear: partition the given unit square S into 7^2 congruent small squares by drawing a 7×7 square grid on S. Since $99 = 7^2 \times 2 + 1$, at least one small square G of the grid contains at least three of the given points. Since the side of G is 1/7, the radius of the circumscribed about G circle is equal to $\frac{1}{7\sqrt{2}}$, and $\frac{1}{7\sqrt{2}} = \frac{1}{\sqrt{98}} < \frac{1}{9}$. Thus, we can surround the square G by a circle Q of radius 1/9. The circle Q is what we are after, for it contain at least 3 given points. ∎

11.4. Observe that $d = a^2 + b^2$ (please see Figure 11.8).

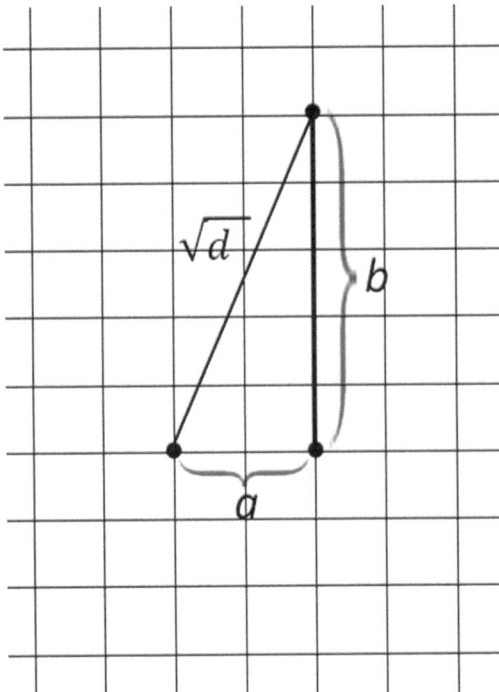

Fig. 11.8

If d is odd, then out of a, b one number is even and one odd, and a chessboard coloring of the vertices of the grid (please see Figure 11.9) insures that any two vertices of the grid distance d apart are colored in different colors.

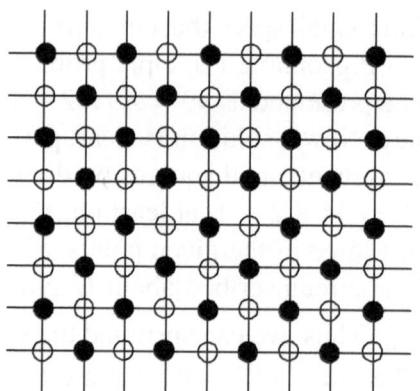

Fig. 11.9

If d is even, i.e., $d = 2d_1$, then should we apply a chessboard coloring to the vertices of the grid (as in Figure 11.9), the distance d would appear *only* between two points of the same color. This observation allows us to partition our coloring problem into two *separate* coloring problems: one for black and another for white vertices, both problems with respect to the distance d_1 (the black vertices form a square grid with the unit $\sqrt{2}$, and so do the white vertices).

If d_1 is even, i.e., $d_1 = 2d_2$, we apply a chessboard coloring to the vertices, and further reduce the problem to coloring two square grids with respect to the distance d_2, and so on. In finitely many steps we will end up with the problem of forbidding a monochromatic odd distance d_n, which we have solved to begin with.

You will find the continuation of this Conway-Soifer train of thought in *Further* Exploration E11. ∎

11.5. It is not obvious – is it? – to tell which of the two problems (A), (B) is easier. In fact, problem (A) is very simple while problem (B) requires a certain amount of ingenuity.

11.5(A). Each placed pair of rooks reduces by 2 the number of rows and the number of columns available for further placing. Since 1994/2 is odd, the first player wins regardless of his strategy.

11.5(B). Again we are watching two numbers, the number R of rows and the number C of columns available for further placing respectively. Each placed pair of rooks reduces R, C by 1 and 2, or by 2 and

1 respectively. This is reminiscent of the changes in coordinates, $1 - 2$ or $2 - 1$, resulting due to a move of a knight (please see Figure 11.10).

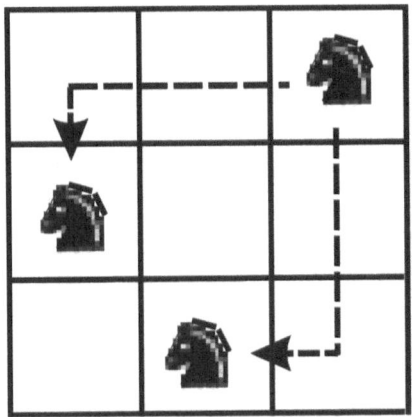

Fig. 11.10

Observe also that one player can keep the knight on a certain diagonal by replying to the reduction $(1, 2)$ by $(2, 1)$ and vice versa (please see Figure 11.11).

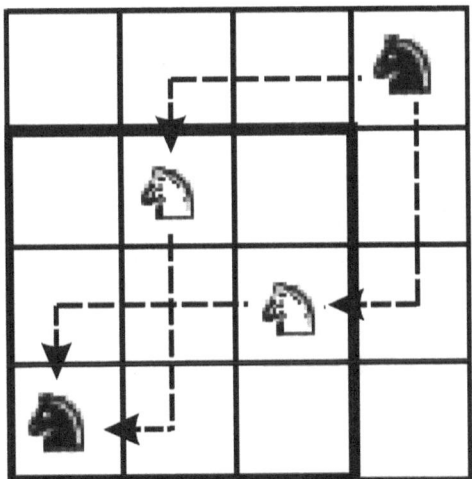

Fig. 11.11

Let (m, n) be a position after a move in our game, where m, n stand for the numbers of available rows and columns respectively after a

player's move. After the first move of the game the position will be
(1992, 1993) or (1993, 1992). Now the winning strategy should be
entering your mind.

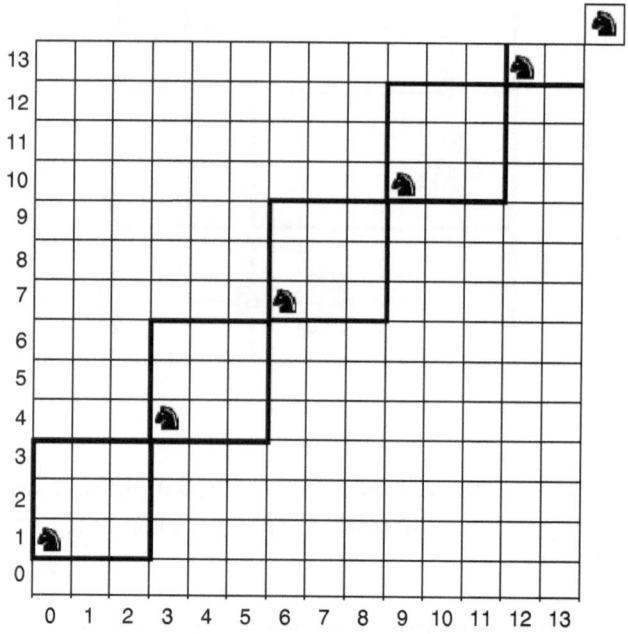

Fig. 11.12

Yes, the first player wins. After his first move, he leaves the position
(1992, 1993). If the second player reduces (R, C) by (2, 1), the first
player reduces (R, C) further by (1, 2) and vice versa, thus always
reducing (R, C) by (3, 3) after the first player's move. Since 1992
is divisible by 3, (1992, 1993) will eventually become (0,1) after the
move of the first player. See Figure 11.12 for a geometric visualization
of the winning strategy. ∎

Twelfth Colorado Mathematical Olympiad
April 28, 1995

Historical Notes 12

Travis Kopp repeats as the winner of the Colorado Mathematical Olympiad

A heavy snow was unable to stop six hundred and three students from coming to the University of Colorado at Colorado Springs for the Twelfth annual Colorado Mathematical Olympiad. Junior high and high school students came from all over the State: Denver, Parker, Fountain, Cañon City, Woodland Park, Florissant, Divide, Englewood, Littleton, Castle Rock, Manitou Springs, Falcon, Peyton, Fort Collins, Hayden, Longmont, Aurora, Franktown, Cascade, Brush, Merino, Hillrose, Ordway, Olney Springs, Erie, Lafayette, Calhan, Yoder, Rush, Sedalia, Larkspur, Monument, Widefield, and Colorado Springs. For the first time in twelve years we also had participants who came from elementary schools.

Participants were offered five problems and four hours of time to solve them and write complete solutions. While Problem 12.2 was solved by over 150 contestants, and Problems 12.1 and 12.3 by 10 and 7 students respectfully, Problem 12.5 was solved by only one participant, Travis Kopp.

Travis became the winner of the Olympiad for the second straight year. Now he was a sophomore from a class of Mrs. Tammy Maculus at Bear Creek High School of Denver.

At the Awards Ceremony, Travis received first prize, which included the gold medal, $1,000 scholarship, a Hewlett-Packard

A. Soifer, *The Colorado Mathematical Olympiad and Further Explorations:*
From the Mountains of Colorado to the Peaks of Mathematics,
DOI 10.1007/978-0-387-75472-7_26, © Alexander Soifer, 2011

graphing calculator with an infrared printer, the book *Colorado Springs – Newport of the Rockies* and the City and Olympiad memorabilia.

Whoever said Mathematics and Beauty cannot coexist! Photo by Tom Kimmel, photographer from *The Gazette Telegraph*

For the second year in a row, second prize was awarded to Scott Mayer, a junior from Fort Collins High School. He received the silver medal, a $300 scholarship, a Hewlett-Packard graphing calculator, a book, and city and Olympiad memorabilia.

Third prizes were awarded to two contestants, both from Palmer High School: Mark Friedberg, a sophomore, and Dan Wright, a junior. Each of them received the bronze medal, a Hewlett-Packard graphing calculator, and city and Olympiad memorabilia.

We also awarded 25 first honorable mentions (two to 6-graders), and 127 second honorable mentions. The Literary award was won by the eighth-grader Stefan Sommars:

Double Header: The Knights Game
By Stefan Sommars

Two players met right over here
While an audience came just to cheer
They played on a board

With spaces galore
A large square with sides of the year.

They looked at the board with a sigh,
One counted and one multiplied.
The number of squares
That they had to share
Three million nine hundred eighty thousand twenty five.

The object was to place a knight
Where the other wouldn't fight
And to find a way
Where the other can't play
Unless he attacks a played knight.

The easiest way you may find
Is to shadow across center lines
Between 997 and 998
Those columns or rows there's a line that is straight
You can mirror him, that is just fine.

If there is a spot he can do
You'll have one just look to and fro,
You can have a play
Until your opponent will say
"There are no more spots, let's go home."

The May 5, 1995, Award Presentation Ceremony included a *Review of the Solutions* by Alexander Soifer and the lecture *Squares in a Square: Investigations by Paul Erdős and Alexander Soifer*, presented by the second author.

The official part included letters from Colorado Governor Roy Romer and Senate Majority Leader Jeffrey M. Welles, and speeches by Kenneth R. Rebman, Vice-Chancellor for Academic Affairs, UCCS; Mary Lou Makepeace, Councilwoman, City of Colorado Springs; John D. Putnam, Past President, Colorado Council of Teachers of Mathematics; Dean James A. Null, College of Letters, Arts and Sciences; and Dean Greg R. Weisenstein, School of Education.

My long letter to the *GazetteTelegraph*, the major newspaper of Colorado Springs, was published in its entirety on May 11, 1995. I have got to include here a short quote:

> The Gazette likes to get the approval of its readership (who does not?). The easiest way to get approval is to bash American secondary education. I would be first to agree that the problems with education are major, and by the year 2000 America will not be No. 1 in the world in math and science education in spite of assurances by Presidents Bush [the father] and Clinton and Gov[ernor of Colorado] Romer. I wish, however, the Gazette would be constructive, to be part of the solution. What can the Gazette do? The Gazette can try to comprehend and accordingly share with the readers that we the people, the state, the society, hold in high esteem achievements in math and sciences. As it now stands, it suffices for a kid to win (or lose) one game in sports, just against another school, to get his or her name in the Gazette. However, to be among the best of the best in the state of Colorado in mathematics in not high enough achievement in the eyes of Gazette editors!

> The Gazette received a list of 157 best students, who were winners of the Olympiad. It published just four names, and even then, as a true provincial paper, the Gazette emphasized bronze medals of two local kids. An astonishing gold medal performance of Travis Kopp did not interest the editors because Travis is a "foreigner" to them: He is a Denverite! GT Editors: The world does not end at County Line Road!

Problems 12

Problems 12.3 and 12.5 for this year's Olympiad were created by my old friend Semjon Slobodnik of Moscow and given to me by the author during my 1988 visit to the land of my birth. Problem 12.4 was created by the legendary Paul Erdős and me on the last day of 1991 when we worked together in my Colorado Springs home on the book of his favorite open problems (my publisher *Springer* hopes I will finish this joint book by 2012). Problems 12.1 and 12.2 were contributed by me. The problems were selected and edited by the Problem Committee: Gary Miller, Jeffrey Smith, Alexander Soifer, and Mark A. Williams.

12.1 KNIGHT MOVES (*A*. Soifer). Two players in turn place knights on a 1995 × 1995 chessboard, one knight per turn, so that a newly placed knight does not attack any knights previously placed on the board. The last player to place a knight wins. Find a strategy that allows one player to win regardless of how another one may play. (A knight *attacks* every square of the board that can be reached by a result of first taking the knight two squares in a horizontal or vertical direction, and then moving it one square in a direction perpendicular to the first direction.)

12.2 TRUE COLORS (*A*. Soifer). Find the minimum number of colors required for coloring unit squares of a 1995 × 1995 chessboard in such a way that any time an L-tromino covers three unit squares of the board, no two of the covered squares have the same color. (An *L-tromino* is a figure that consists of three unit squares of the board that share a vertex.)

12.3 DON'T COME TOO CLOSE! (*S. G. Slobodnik*). A straight line L is drawn through two vertices of an infinite square grid. Prove that there is a number $a > 0$ such that no vertex of the grid not on L lies at a distance less than a from L.

12.4 WHAT'S THE POINT? (*P. Erdős and A. Soifer, December 31, 1991*). Given a triangle T, for every point p inside T we can compute the product xyz of the distances x, y, and z from p to the three sides of T. Find the point p at which the product xyz reaches a maximum.

12.5 COVER-UP (*S. G. Slobodnik*). A square grid with cells of size 1 × 1 inches is tiled by polygons, each of area 1.5 square inches. Prove that if there are no grid vertices on the perimeters of the polygons and all polygons contain an equal number of grid vertices, then there is a polygon whose diameter is greater than 10 inches. (We say that a grid is *tiled* by polygons if the grid is completely covered by polygons, with no two polygons having an inside point in common. The *diameter* of a polygon is the greatest distance between two of its points.)

Solutions 12

12.1. The first player has a winning strategy. On his first move he places a knight in the central square of the board, and then places his knights symmetrically to the knights placed by the second player with respect to the center of the board (please see Figure 12.1). ■

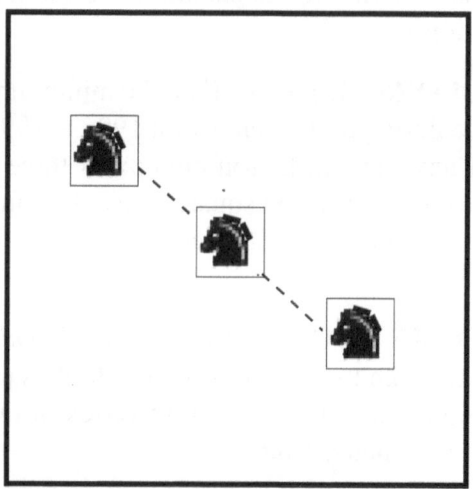

Fig. 12.1

12.2. The 2 × 2 grid must be colored in at least four colors, for otherwise an L-tromino can be placed on it to cover more than one square of a color (please see Figure 12.2). Thus, four colors are needed.

4	1
3	2

Fig. 12.2

We can now cover the entire 1995 × 1995 chessboard by disjoint 2×2 squares (please see Figure 12.3), each 4-colored as in Figure 12.2

(some 2×2 squares can stick out of the given chessboard, which is all right). This 4-coloring shows that four colors also suffice. ■

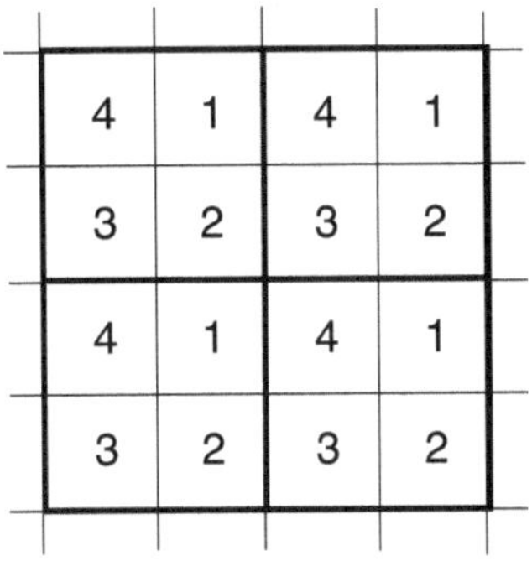

4	1	4	1
3	2	3	2
4	1	4	1
3	2	3	2

Fig. 12.3

12.3. Let e be the line through (at least) two vertices of the square grid, and let M_1 and M_2 be the vertices nearest to each other on e. Let A be a vertex of the grid not on e. We construct a parallelogram $ABCD$ such that A and B are symmetric to each other with respect to M_1, and C and D are symmetric to each other with respect to M_2 (please see Figure 12.4).

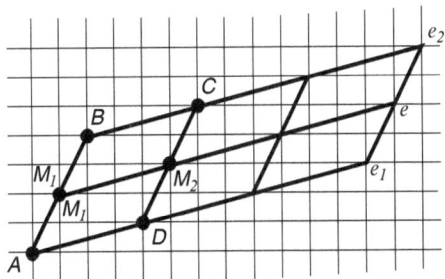

Fig. 12.4

Since there are only finitely many vertices of the grid inside or on the boundary of the parallelogram $ABCD$, there is a vertex B of the grid that is at the minimum distance a from e. Now we can cover the entire line e by the strip of translates of the parallelogram $ABCD$.

In this strip the minimum distance of a grid vertex from e is exactly the same as in $ABCD$. Therefore there is no vertex of the grid closer to e than a. ∎

12.4. Let the side lengths of the triangle T be a, b, c respectively. Pick a point O arbitrarily inside T, and drop perpendiculars to all sides of T; let their lengths be x, y, z respectively (please see Figure 12.5).

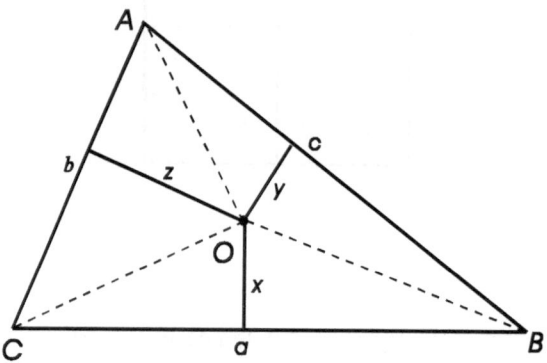

Fig. 12.5

Then the broken lines partition T into three triangles, and we get the following equality of areas, where S denotes the area of T:

$$xa + yc + zb = 2S.$$

You can easily prove (please do), that the product of three numbers whose sum is constant reaches a maximum when the numbers are equal. Consequently, the previous equality implies that the maximum of $xa \times yc \times zb = abcxyz$ is reached when $xa = yc = zb$. Since a, b, c are constant side lengths, the desired maximum of xyz is attained at exactly the same point, i.e., when $xa = yc = zb$. But this equality defines O as the center of mass of the triangle T!

If you were unable to prove that maximum of the product of 3 numbers, whose sum is a positive constant, reaches maximum when the numbers are equal, I can show it to you now. Assume that $a + b + c = 3n$ for a constant n and let $a \leq n \leq b$. Then you can verify that $(a + b - n)n - ab = (b - n)(n - a) \geq 0$, which implies that $abc \leq n(a + b - n)c$. Similarly you can prove that $n(a + b - n)c \leq nnn$. The last two inequalities produce the desired inequality $abc \leq nnn$. ■

12.5. We will argue by contradiction: assume that all the conditions of the problem have been satisfied, and the diameter of each polygon of the tiling is less than 10 inches. Let us take a look at the 1000×1000 inches square Q with corners in the vertices of the grid and sides parallel to the gridlines. Let each of the tiling polygons contain k vertices of the grid. Then the number of polygons having non-zero intersection with the square Q is greater than $\frac{1000^2}{k}$ and less than $\frac{1021^2}{k}$. The sum of the areas of these polygons is greater than 1000^2 and smaller than 1020^2. These two considerations imply that the area S of each polygon satisfies the following double inequality:

$$\frac{1000^2}{1021^2} \cdot k < S < \frac{1020^2}{1000^2} \cdot k.$$

But $S = 1.5$, and from the above we get two inequalities in k:

$$1.5 < \frac{1020^2}{1000^2} \cdot k, \text{ therefore, } k > 1;$$

and

$$\frac{1000^2}{1021^2} \cdot k < 1.5, \text{ therefore, } k < 2.$$

For a whole number k, it is "difficult" to be strictly between 1 and 2. We have arrived at a contradiction that proves the desired result. ■

Thirteenth Colorado Mathematical Olympiad
April 19, 1996

Historical Notes 13

Top young Colorado mathematician wins a computer made by Apple people of Colorado

The thirteenth Annual Colorado Mathematical Olympiad brought together some 600 junior high and high school students. Contestants came from all over Colorado: Denver, Parker, Canon City, Woodland Park, Littleton, Castle Rock, Fort Collins, Manitou Springs, Castle Rock, Rangely, Ellicott, Durango, Longmont, Foxfield, Erie, Franktown, Sanford, Falcon, Calhan, Fountain, Aurora and Colorado Springs.

The winner of this year's Olympiad was Steve Schell, a junior from the class of John Wicklund at Doherty High School. At the April 26, 1996, Award Presentation Ceremony Steve received first prize, which included the gold medal, Power Macintosh Computer system made here in Fountain, Colorado, and donated by Apple Computer, and the book *Colorado Mathematical Olympiad: The First Ten Years and Further Explorations* by Alexander Soifer.

Second prizes were awarded to two contestants: Daniel Wright, a senior from Palmer High School; and Linda Lin from Fort Collins High School. Each of them received the silver medal, a $650 scholarship, Texas Instruments TI-92 super calculator, and the Olympiad book by A. Soifer. Linda also received a rare collector's Cray Computer paperweight with a gallium chip inside. You may have heard of the incredible inventor Seymour Cray, the creator of the supercomputer. He left Cray Research, a company he founded, and started Cray

A. Soifer, *The Colorado Mathematical Olympiad and Further Explorations:* *From the Mountains of Colorado to the Peaks of Mathematics*, DOI 10.1007/978-0-387-75472-7_27, © Alexander Soifer, 2011

Computer company right here in Colorado Springs! We presented it to Linda as the only contestant to substantially advance in the solution of the most difficult problem" So Close, Faraway!" So much for the allegation that girls do not make fine mathematicians!

Third prizes were presented to the following four participants: Andy Hoke, a junior, Air Academy High School; Nick Sanford, a sophomore, Fountain-Fort Carson High School; Jerry Tolzman, a senior, Manitou Springs High School; and Forrest Brinker, a junior, Ponderosa High School. Each of them received the bronze medal of the Olympiad, top of the line Hewlett-Packard 48G calculator and the Olympiad book by A. Soifer.

We also awarded 3 fourth prizes, 32 first honorable mentions, and 70 second honorable mentions. This year we had a guest contestant Tarik Kabil from Bosnia, who won an Appreciation of the Judges certificate.

The Award Presentation program included *Review of Solutions* by Alexander Soifer, lectures *Making Friends and Learning Graph Theory* by Kenneth R. Rebman, and *Adventures in Map Coloring* by Alexander Soifer.

The following guests of honor, hosts, and sponsors addressed the winners and present the awards: Jeffrey M. Wells, Colorado State Senate Majority Leader; Greg Hoffman, Human Resources Manager, Apple Computer; John D. Putnam, Past President, Colorado Council of Teachers of Mathematics; and the following leaders of UCCS: Kenneth R. Rebman, Vice-Chancellor for Academic Affairs; James A. Null, Associate Vice-Chancellor for Academic Affairs; Greg R. Weisenstein, Dean, School of Education; and Douglas E. Swartzendruber, Dean, College of Letters, Arts and Sciences.

The winners were addressed, via an April 26, 1996, letter, by the Governor of the State of Colorado Roy Romer:

> On behalf of the citizens of Colorado, I would like to extend greetings and congratulations to the students and their parents and teachers who are being honored at the 13th Colorado mathematical Olympiad.
>
> As you may know, I have devoted a significant portion of my three terms as Governor to the promotion and enhancement of our educational system in Colorado. I believe that our future as a state and nation depends on a strong foundation in education. It gives me a great pleasure to see an event that highlights those who excelled academically in mathematics.

This year, the Mathematical Association of America appointed me to my first three-year term on the six-member United States of America Mathematical Olympiad (USAMO) Subcommittee, which created, selected, and edited problems for USAMO, and graded papers of 100-200 participants. I ended up serving USAMO from 1996 to 2005. In order to get to USAMO, students had to excel in the American Mathematics Contest (multiple choice) and the American Invitational Mathematics Examination (answer only). This requirement of succeeding though multiple choice and answer-only contests in order to have the right to write complete solutions always looked ridiculous to me. It is like asking contestants in piano competition to first sing and dance! We invite everyone to come and write solutions at the Colorado Mathematical Olympiad!

This year in Seville, Spain, I was also elected secretary and member of the Executive of the World Federation of National Mathematics Competitions (WFNMC). I was reelected in 2000 in Tokyo, Japan, and in 2004 in Copenhagen, Denmark. In 2008 members elected me to a four-year term as Senior Vice President of WFNMC.

Problems 13

This was the only year when we "dressed" all five Olympiad problems as games and puzzles. Problems 13.1, 13.2, 13.3 (the "minimum" part of it) and 13.4 were created by me especially for this year's Olympiad (the "maximum" part of 13.4 came from the mathematical folklore). Problem 13.5 came from the unpublished problems of the 1988 preparation of the Soviet Union Team for the International Mathematics Olympiad. Many years later I learned from my Latvian friend Professor Agnis Angans that this problem (13.5) was created by Aivars Bērziņš, Associate Professor at the University of Latvia. Amazingly, I had met Aivars in the distant past. He was a young slim undergraduate student (I was a Ph.D. student at the time); we shared a room as judges of the Soviet Union National Mathematical Olympiad in Kishinev (now the capital of Moldova) in May 1973, some 37 years ago!

The problems were selected and edited by the Problem Committee: Gary Miller, Ed Pegg, Jr., Alexander Soifer, and Mark A. Williams.

Fun and Games '96

13.1 ROCKING GAME (*A. Soifer*). Three piles of rocks contain 1995, 1996, and 1997 rocks respectively. Two players in turn take any number of rocks from any one or two piles (they must take at least one rock from at least one pile). The player who takes the last rock wins. Find a strategy that allows one player to win regardless of how another one may play.

13.2 MOVIN' ALONG THE POLYGON (*A. Soifer*). The first of two players puts his white piece and the opponent's black piece on two distinct vertices of a given regular 1996-gon. He then moves the white piece to an unoccupied vertex and draws the straight line from the old to the new position of his piece. The second player then moves the black piece to a currently unoccupied vertex of the 1996-gon and draws the line his piece thus traveled, etc. One is not allowed to move along previously drawn lines. The player who makes the last move wins. Find a strategy that allows one player to win regardless of how another one may play.

13.3 HOW FEW ARE TOO FEW? HOW MANY ARE TOO MANY? (*A. Soifer*). Find the minimum and the maximum numbers of kings that can be placed on a 1996×1996 board so that every square of the board is attacked or occupied, and no kings attack each other. (A king attacks all squares that have at least one point in common with the square occupied by the king.)

13.4 GRIDDY COLORING GAME (*A. Soifer*). Given a 1996×1996 square grid. Two players in turn color red one previously uncolored unit edge of the grid (boundary edges are included). The player who creates the first closed red walk loses. Find a strategy that allows one player to win regardless of how another one may play.

13.5 SO CLOSE, FARAWAY! (*A. Bērziņš*). Some unit edges of an infinite square grid are colored red so that one can travel along red edges from any vertex of the grid to any other vertex along red edges, but there are no closed red cycles[1]. Prove that there are two neighboring vertices in the grid such that the shortest walk along red edges between them is greater than 1996 unit edges long.

Solutions 13

13.1. The first player wins. With his first move, the first player takes one rock from the second pile and two rocks from the third one, thus leaving the piles with 1995 rocks each. No matter how the second player may play, the first player can always *equalize numbers of rocks in all of the three piles*, thus making them eventually all equal to zero after his last move, which includes taking the last rock. ∎

13.2. *First Solution.* Let the starting positions of the black and the white pieces be B_1 and W_1 respectively. The second (black) player wins by making his moves symmetrically to the moves of the first (white) player with respect to the line with respect to which B_1 and W_1 are symmetric (please see Figure 13.1).

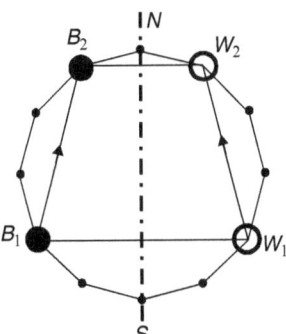

Fig. 13.1

Since 1996 is even, if the symmetry line contains a vertex of the given 1996-gon, then it contains another vertex of it (N and S in Figure 13.1). In this case, if the first player moves the white piece

---

[1] We say that a walk along red edges is a *closed red cycle* if no edges are repeated along the walk and it ends at the vertex it started.

to one of the vertices N, S, the second player moves the black piece to the other of these vertices. ∎

Second Solution. I noticed that "the heart" of this problem is not a geometric symmetry that was used in the first solution, but a *set-theoretic symmetry*!

Fix a one-to-one correspondence between one half of the 1996-gon vertices and the other half, in which B_1 corresponds to W_1. When the first player moves the white piece to a vertex, the second player moves the black piece to the corresponding vertex. That's all! ∎

13.3.

(a) *Maximum.* Split the board into 2×2 cells. Each cell can contain at most one king, for two kings in the same cell would attack each other. Thus, the board can contain at most as many kings as there are 2×2 cells, i.e., $\left(\frac{1996}{2}\right)^2 = 998^2$ kings.

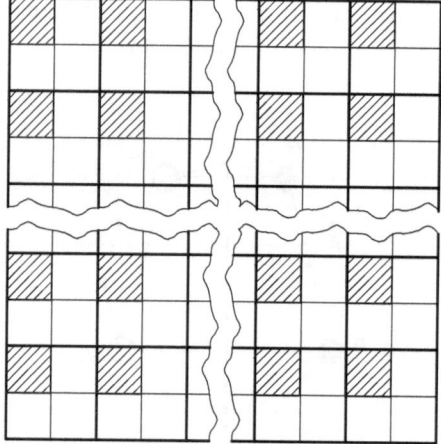

Fig. 13.2

Figure 13.2 shows that you can actually place this many kings in the striped squares of the board. ∎

(b) *Minimum.* Split the board into 3×3 cells with 2×3 cells on the boundaries and 2×2 cells in the corners, and In each cell mark by stripes one square (see Figure 13.3). Observe that a striped square cannot be attacked by any king, located outside of that cell. Thus, the board must contain at least as many kings as there are cells in our partition of the board, i.e., at least $\left(\frac{1998}{3}\right)^2 = 666^2$ kings.

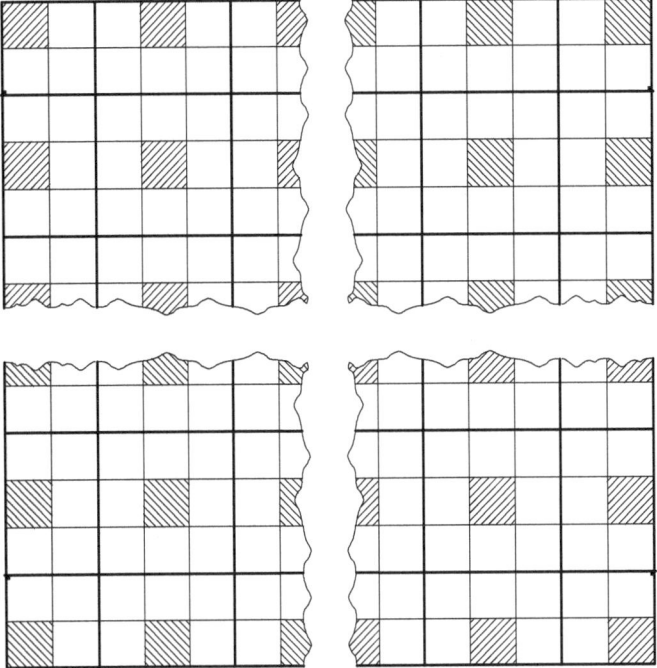

Fig. 13.3

Figure 13.3 shows that 666^2 kings placed on the striped squares of the board indeed attack or occupy every square of the board. ■

13.4. It makes sense to consider the notion of a *subgrid* Q' of the given 1996×1996 square grid Q to mean a subset of the vertex set of Q some of which are connected by (unit) edges from Q. We will call a subgrid Q' a *tree* if it contains no cycles (please see Figure 13.4).[2]

A tree Q' is called *spanning tree* if adding any single edge (and incident to this edge vertices, if they are not yet in the vertex set of Q') would make it a non-tree, i.e., would create a cycle. In this convenient terminology, we can pose natural relevant questions:

Does each spanning tree of Q contain all of the vertices of Q?

---

[2] We say that a walk along edges is a *cycle* if no edges are repeated along the walk and it ends at the vertex where it started.

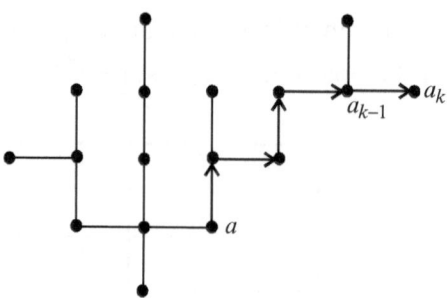

Fig. 13.4 A tree

The answer is yes. Indeed, if a vertex q is not a part of the spanning tree Q', none of the edges emanating from q is a part of the edge set of Q'. Let e be one such edge emanating from q. Adding e to the edges of Q' and both endpoints of e to the vertices of Q' would not produce a closed walk, and thus would contradict the maximality of Q'.

Do all spanning trees of Q have the same number E of edges?

The answer is yes, moreover, the number of edges E in any tree Q' is one less than its number of vertices P:

$$E = P - 1. \tag{*}$$

We can easily prove the equality (*) by induction on the number P of vertices. For $P = 1$, we have $E = 0$, and the equality (*) holds.

Assume that the equality (*) is true for trees with P vertices, where $P > 1$. Let now Q' be a tree with $P + 1$ vertices. Pick a vertex in Q', any vertex, and start traveling along the edges of Q'. We will never visit the same vertex a second time, for this would create a closed walk, impermissible in a tree. Therefore, we will eventually get into a "dead end," a vertex p of degree 1 (i.e., exactly one edge e emanates from p, and we arrived in p by traveling along e). We now remove from Q' the edge e and the vertex p, and call the resulting tree Q". Obviously, Q" has P vertices, and thus by the inductive assumption, Q" satisfies the equality (*): $E = P - 1$. This implies that $(E + 1) = (P + 1) - 1$. The last equality means exactly that the tree Q' satisfies the equality (*) too.

With all the prep job done, it is very easy to find the winning strategy for our game. Since the total number of vertices in the

1996×1996 square grid Q is 1997^2, the number of edges in *any* spanning tree Q' of Q is $1997^2 - 1$, an *even number*. Therefore, the second player wins. He has merely to avoid creating a cycle – and otherwise play in any way he pleases! If the first player too were to avoid creating a cycle as long as he could, the players will eventually create a spanning tree. Since the number of edges in this spanning tree will be even, the second player will make the last safe move. ■

13.5. First observe that for any two vertices A and B of the grid, not only does a path AB exist, but it is unique, for otherwise the union of two distinct paths AB would contain a red cycle.

Now let us pick two vertices A and B of the grid that are distance 2000 unit edges apart (in the sense of the usual Euclidean distance), and connect them by the red path AB (which zigzags from A to B; see Figure 13.5). On this red path there is a vertex O that is distance at least 1000 away from A and from B. Let T denote a vertex on the red path AB that is adjacent to O. We draw a circle C of radius 999 and center at O. The vertices A and B obviously lie outside the circle C. Let us now color all vertices of the grid that lie outside C in two colors, yellow and blue, as follows:

We color a point P yellow if a red path OP does not pass through the point T, and blue if it does. For example, we color B yellow and color A blue (Figure 13.5).

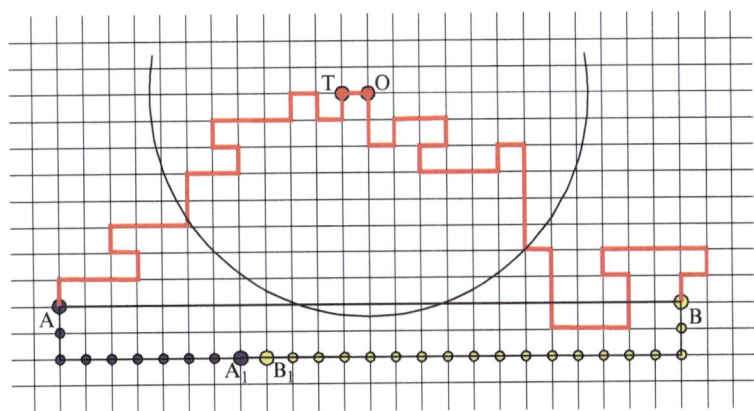

Fig. 13.5

Let us now connect the vertices A and B by a path along the lines of the grid that lies entirely outside the circle C (you can see it in the lower part of Figure 13.5). As we travel from the blue vertex A to the yellow vertex B we will positively encounter at least one (maybe more than one!) switch of color: let this switch happen from the blue vertex A_1 to the yellow vertex B_1. We can now easily prove that the red path from A_1 to B_1 is longer than 1996 unit edges. Indeed, we can assemble a red path from A_1 to B_1 by combining the red path A_1TO with the red path OB_1, each of which is surely at least 1000 unit edges long. Observe, this A_1TOB_1 is the only red path that connects A_1 to B_1. We are done! ■

Fourteenth Colorado Mathematical Olympiad
April 25, 1997

Historical Notes 14

Doherty's Steve Schell repeats as Colorado's best young mathematician in the Olympiad postponed by an April "avalanche"

The Olympiad season commenced with the April 17, 1998, *Gazette* article "Calling all creative types to state Math Olympiad" by Wendy Y. Lawton:

Number geeks need to apply. Calculus fiends, geometry nerds, algebra wonks – forget it.

This is a cattle call for creative types.

To be a contender at the Colorado math Olympiad, coming up next week, kids need some imagination. They need a touch of a poet. They need, to use the current corporate cliché, to "think outside the box."

Just ask Alexander Soifer, a professor of mathematics at the University of Colorado at Colorado Springs, sponsor of the competition for middle and high school students since 1983.

"The students who do well at the Olympiad have freedom of thought," Soifer said. "They can look at something in a new way, which is true of painters."

Painters doing math?

That may not be far-fetched, given the type of math mind-benders offered in the contest. There are few formulas, equations or intricate calculations involved. In fact, kids can bring a calculator.

What the Olympiad is about is solving problems – five word problems in four hours to be exact. While a solid grasp of math concepts

A. Soifer, *The Colorado Mathematical Olympiad and Further Explorations: From the Mountains of Colorado to the Peaks of Mathematics*, DOI 10.1007/978-0-387-75472-7_28, © Alexander Soifer, 2011

will come in handy to grab one of the top prizes, such as a personal computer or some scholarship money, what is really needed is an open, analytical mind...

This focus on using both sides of the brain is fitting. Albert Einstein played the violin. Pythagoras dabbled in religion. Heck, a past president of the American Mathematical Society, is an avid juggler.[1]

And so it goes with Olympiad winners, who are athletes, musicians and drama buffs. One recent winner was apparently a hunk. When he received his prize, he was surrounded by pretty girls.

"All our winners are very much alive," Soifer said. "They are not narrow people."

For the first time in its 14-year history, the Colorado Mathematical Olympiad was buried by two feet of snow, so much for the spring, the April 25th date! At 5 in the morning I was awakened and informed by the campus police that the University would be closed, and I had to postpone the Olympiad by a week. However, at 8 in the morning I received a phone call from the campus police again. This time they informed me that 10 students had arrived to participate! What was I to do?

I jumped in the car and drove to the campus as swiftly as the snowy roads allowed. There I found Mel Oliver, a fine mathematics teacher from Rangely High School, and his nine students, and Michelle Bohren from Durango with her mom. Both groups had traveled many hours to make it to Colorado Springs. I invited them all to one of Colorado Springs' finest restaurants, *Giuseppe's Depot*, for brunch. It was a great opportunity for these 10 students, their teacher, and I to have a mathematical conversation, without any rush, with the snowed city as a backdrop outside the restaurant's windows. In consideration of their circa 15 hour roundtrip, I also made one and only exception in the history of the Olympiad: I allowed these 10 students to compete in the Olympiad in their own schools, proctored by my absolutely trusted colleagues Mel Oliver in Rangely, and Professor Dick Gibbs in Durango.

A week later, on May 2, 1997, the Fourteenth annual Colorado Mathematical Olympiad had finally taken place. It brought together

---

[1] The author clearly refers to the great mathematician and my friend Ron Graham.

427 middle and high school students (an amazingly high attendance considering the postponement). Contestants came from all over Colorado: Denver, Thornton, Parker, Woodland Park, Littleton, Castle Rock, Fort Collins, Manitou Springs, Castle Rock, Ellicott, Longmont, Erie, Falcon, Calhan, Fountain, and Colorado Springs. In addition, as you know, the Olympiad was administered in Rangely and Durango to 10 students, who had braved the snow and come to Colorado Springs on the original day of the contest.

Steven Schell, a senior from the class of John Wicklund at Doherty High School won first prize for the second year in a row. At the May 17, 1997 Award Presentation Ceremony Steve received the gold medal of the Olympiad, a Power Macintosh Computer made here in Fountain by SCI Systems, and donated by SCI and Apple Computer, and the 1994 book *Colorado Mathematical Olympiad: The First Ten Years and Further Explorations* by Alexander Soifer.

Second prize was awarded to Aaron Parsons, a junior from Rangely High School. Aaron was one of those who braved the snow on the original day of the contest. He received the silver medal, a $600 scholarship, a Hewlett-Packard 48G calculator with infrared printer, and the first 1994 version of this Olympiad book by A. Soifer. Do not forget about Aaron – you will meet him again in the next installment of *Historical Notes* and in Part V!

Third prizes were presented to the following two participants: John Batchelder, a sophomore from Columbine High School in Littleton; and John Crockett, a sophomore from Doherty High School. Each of them received the bronze medal of the Olympiad, a $300 scholarship, Hewlett-Packard 48G calculator, and the 1994 Olympiad book by A. Soifer. Do not forget these two Johns, Batchelder and Crockett – you will meet them again next year!

We also awarded 2 fourth prizes, 19 first honorable mentions, and 16 second honorable mentions. The Literary award was presented to Claire Darby, from Air Academy High School.

The prize fund of the Olympiad had been generously donated by SCI Systems , Apple Computer, Hewlett-Packard, Casio, Texas Instruments, CU-Colorado Springs Bookstore, Colorado Springs School District 11, Air Academy School District 20 Harrison School District 2, Fort Collins High School, St. Mary's High School, and Colorado College.

The Award Presentation program included *Review of Solutions* by Alexander Soifer and the lecture *In Memory of Paul Erdős* by Alexander Soifer. The following guests of honor, hosts and sponsors addressed the winners and present the awards: Linda Bunnell-Shade, Chancellor of UCCS; Elizabeth S. Grobsmith, Dean, College of Letters, Arts and Sciences; Jeffrey M. Wells, Colorado State Senate Majority Leader; James A. Null, Colorado Springs Councilman; Greg Hoffman, Personnel Manager, SCI Systems, Inc.; and Kathleen Pool, Supplier Quality Engineer, Apple Computer, Inc.

Problems 14

Problem14.1 was created for this year's Olympiad by Alexander Soifer. The theme of celebrated cartoons about Fred Flintstone and Barney Rubble was suggested by Maya Soifer; we will return to this theme many times in the future. Problem 14.2 was adapted by A. Soifer from the Russian mathematical folklore. Problem 14.3 was created for this Olympiad by Dr. Semjon Slobodnik of Moscow, Russia. In view of Paul Erdős's passing, I wished to use one of his results in this year's Olympiad. I chose Problem 14.4 from Paul Erdős's 1946 paper "On Sets of Distances of *n* Points" [E0]. Problem 14.5(b) was used in 1990 in the selection of the Soviet National Team for the International Mathematics Olympiad; I strengthened it in 1995, thus creating problem 14.5(c). The problems were selected and edited by the Problem Committee: Robert Ewell, Gary Miller, and Alexander Soifer.

This was the first Olympiad following the passing of Paul Erdős. I dedicated it to this great mathematician, teacher, coauthor, and dear friend:

In Memory of Paul Erdős, 1913–1996

14.1 STONE-AGE ENTERTAINMENT (*A. Soifer*).
A pile contains 99 pebbles. Fred Flintstone and Barney Rubble in turn take pebbles from the pile: first Fred takes 1 pebble; then Barney takes 1 or 2 pebbles; then Fred takes 1 or 2 or 3 pebbles; then Barney takes 1 or 2 or 3 or 4 pebbles; etc. The player who takes the last pebble wins. Find a strategy that allows Fred or Barney to win regardless of how the other one may play.

14.2 1997 SQUARED. Every unit square of a 1997×1997 board is occupied by $+1$ or -1. For every row i we compute the product R_i of all numbers in this row; similarly for every column j we compute the product C_j of all numbers in this column.

Prove that the sum $S = R_1 + R_2 + \ldots + R_{1997} + C_1 + C_2 + \ldots + C_{1997}$ cannot be equal to zero.

14.3 CUBE IN THE BOX (*S. G. Slobodnik*). A toy factory manufactures cubes and packages each of them in a box with a cover. Each cube is colored in six colors: each face in one color, and all faces in different colors. Each packing box with the cover is also colored in the same six colors: each face in one color, and all faces in different colors (but not necessarily in the same way as the cube). Isabelle, who received this Cube in the Box for her birthday, is trying to put the cube in the box in such a way that the touching faces of the cube and the box all differ in color. Can Isabelle always succeed?

14.4 THE ERDŐS PROBLEM (*P. Erdős*, 1946, [E0]). Given a positive integer n. Let the maximum and minimum distances determined by n points in the plane be denoted by R and r respectively.

(A) Prove that in any set of n points in the plane r can occur at most $3n$ times.

(B) Prove that in any set of n points in the plane R can occur at most n times.

14.5 SQUARES IN A SQUARE (*A. Soifer*).

(A) Prove that any two squares whose areas add up to 1 can be inscribed with no interior points in common in a square of area 2.

(B) Prove that any four squares whose areas add up to 1 can be inscribed with no interior points in common in a square of area 2.

(C) Prove that any five squares whose areas add up to 1 can be inscribed with no interior points in common in a square of area 2.

Solutions 14

14.1. Behold:
$$3 + 5 + \ldots + 19 = 99.$$

This equality immediately gives a winning strategy for Barney Rubble: he takes the difference between $3, 5, \ldots, 19$ and the number of pebbles taken by Fred Flintstone respectively. Barney's strategy allows for a nice geometric illustration (Figure 14.1). Barney puts the pebbles taken by Fred into consecutive L-shape layers and takes all remaining pebbles from that "L"! ■

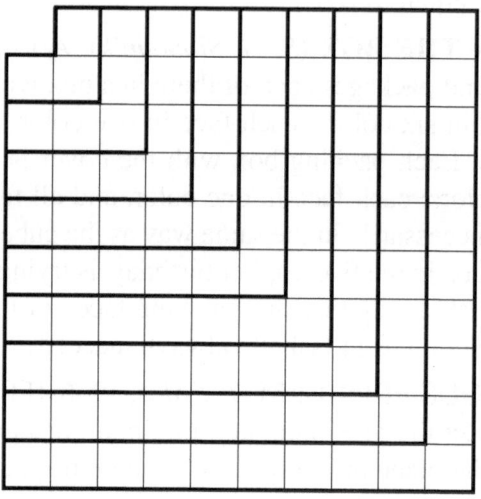

Fig. 14.1

The problem allowed a nice solution for the pile of 99 pebbles. What would happen if the pile had, say, 100 pebbles? This and other related problems comprise *Further Exploration* E12: *Stone Age Entertainment* later in this book!

14.2. If we switched the sign of one number located in the intersection of row i column j (Figure 14.2), we would change the sign of exactly two summands R_i and C_j and thus would change the sum $S = R_1 + R_2 + \ldots + R_{1997} + C_1 + C_2 + \ldots + C_{1997}$ by $\pm 2 \pm 2$, i.e., by 0 or ± 4 – in other words, by a factor of 4.

Assume that the sum S is zero. Change signs of each negative number, one at a time, until we get all positive 1's. At every change the sum S would change by a multiple of 4, and therefore, the final sum would be a multiple of 4. On the other hand, the sum for the board filled with 1 in each square, is obviously $1997 + 1997$, which is not a multiple of 4, a contradiction.

We have thus proved that the sum S cannot be equal to zero no matter how the board is filled with 1's and (-1)'s. ■

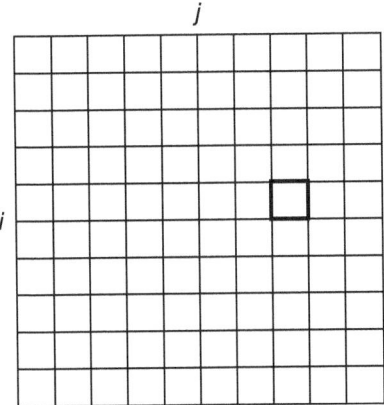

Fig. 14.2

14.3. Denote the colors of one pair of opposite faces of the box by 1 and 2, and the remaining colors by 3, 4, 5, and 6. Out of three pairs of opposite faces of the cube, at least one pair does not use colors 1 and 2; let the colors of this pair be 3 and 4. Now place the cube in the box so that the cube's face of color 3 faces the box's face of color 1 (then of course, the cube's face of color 4 faces the box's face of color 2).

The remaining box's faces are colored 3, 4, 5, 6, while the remaining cube's faces are colored 1, 2, 5, 6; these 4-element sets share only two colors (5 and 6). On the other hand, there are 4 rotations of the cube about the axis through its faces colored 3 and 4 please see Figure 14.3).

Therefore, we can rotate the cube to a position where faces of the box and the cube of colors 5 and 6 do not coincide. ∎

14.4.

(A) Observe that there are at most 6 minimum distances r emanating from a point (see point V_1 in Figure 14.4), for every angle $\angle V_2 V_1 V_3 \leq \frac{\pi}{3}$. Since n points are given, we get at most $6n$ distances r. Each distance here was counted twice – once at each of its endpoints – therefore there are at most $3n$ distances r in any configuration of n points.

(B) Observe that any two segments of the maximum length R must intersect. Assume that they do not, i.e., $|AB| = |CD| = R$ and AB and CD do not intersect. Then the convex hull of A, B, C, D is either a quadrilateral (Figure 14.5) or a triangle (Figure 14.6).

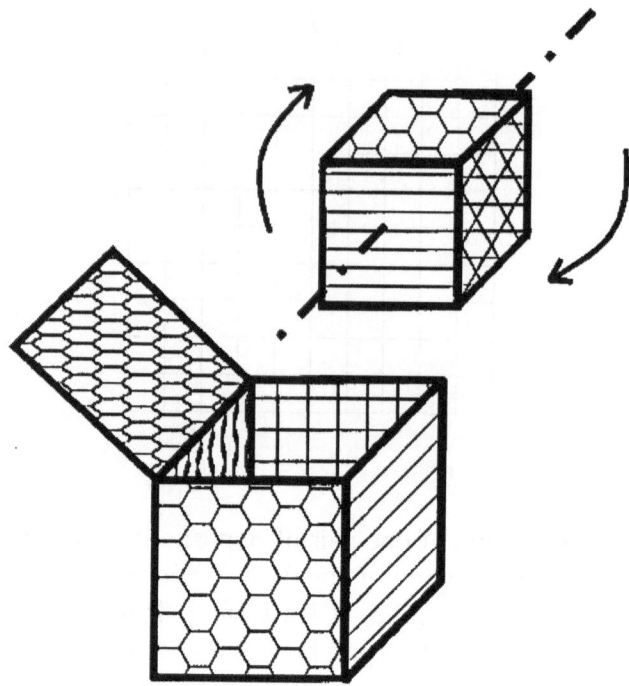

Fig. 14.3

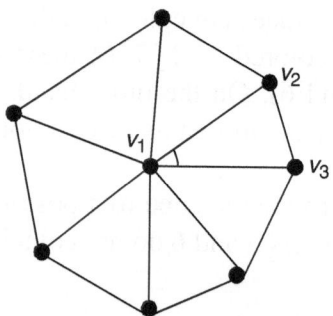

Fig. 14.4

In the first case, one of the angles of the quadrilateral, say, the angle B is obtuse, and we get $|AC| > |AB|$, i.e., we get a longer distance than the maximum, a contradiction. In the latter case, we get $|CD| \leq |CE| < \max(|CA|, |CB|)$, i.e., CA or CB are longer than the maximum, a contradiction again.

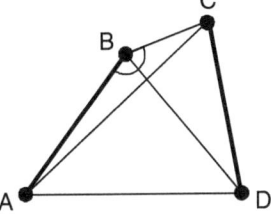

Fig. 14.5

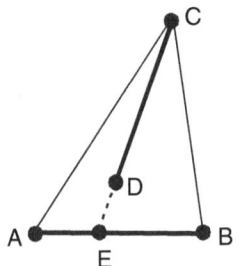

Fig. 14.6

Now we can easily finish, by induction on the number n of points, the proof that there are at most n maximum distances R in any configuration of n points.

For $n = 1$ we obviously have no more than 1 maximum distance (we actually have none!), and the desired statement holds.

Assume that the statement is true for any configuration of $n = k$ points.

Given a configuration of $n = k+1$ points. If at most two maximum distances emanate from any point, we get at most $2(k+1)$ maximum distances, where each such distance was counted twice, for each of its endpoints. So we must divide this upper bound by 2 and get the desired result: we have at most $k + 1$ maximum distances.

Assume now that there are at least three maximum distances R emanating from a point (see point V_1 in Figure 14.7). Then $V_3 V_1$ is the only maximum distance segment emanating from V_3, for any other such segment $V_3 W$ of length R would have to intersect both $V_2 V_1$ and $V_4 V_1$, which is impossible.

But since only one maximum distance emanates from V_3, we remove this one point (and thus eliminate precisely one maximum distance). By the inductive assumption we now have at most k maxi-

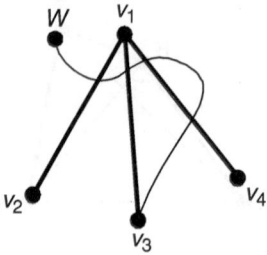

Fig. 14.7

mum distances. Now, by adding V_3 back, we conclude that the given configuration of $k + 1$ points generates at most $k + 1$ distances R.

Come to *The Erdős Train Station*, Further Exploration E13, for more of these problems. ∎

14.5.

(A) Given two squares of sides a_1 and a_2 whose areas add up to 1, we get the equality $a_1^2 + a_2^2 = 1$. By adding this equality with the standard inequality $2a_1a_2 \leq a_1^2 + a_2^2$, we get $(a_1 + a_2)^2 \leq 2$, or $a_1 + a_2 \leq \sqrt{2}$. But this means precisely that the two given squares fit in a square of side $\sqrt{2}$, or area 2; see Figure 14.8.

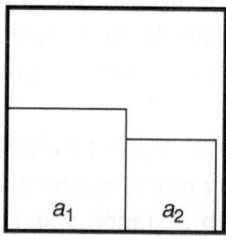

Fig. 14.8

(B) Given four squares of sides a_1, a_2, a_3 and a_4; $a_1 \geq a_2 \geq a_3 \geq a_4$, whose areas add up to 1, we get the equality $a_1^2 + a_2^2 + a_3^2 + a_4^2 = 1$. Just as in (a) above, we can show that $a_1 + a_2 \leq \sqrt{2}$; surely we also get $a_1 + a_3 \leq \sqrt{2}$, and the four squares fit within the square of area 2 as shown in Figure 14.9.

(C) Finally, given five squares of sides a_1, a_2, a_3, a_4, and a_5; $a_1 \geq a_2 \geq a_3 \geq a_4 \geq a_5$, whose areas add up to 1, we get the equality

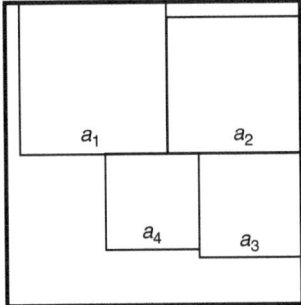

Fig. 14.9

$a_1^2 + a_2^2 + a_3^2 + a_4^2 + a_5^2 = 1$. You will immediately see that the following inequality holds for the three smallest smaller squares: $a_3^2 + a_4^2 + a_5^2 \le \frac{1}{5} + \frac{1}{5} + \frac{1}{5}$. We are now ready for the final assault (we use here the same standard inequality as in part (a) above, but three times):

$$(a_3 + a_4 + a_5)^2 = a_3^2 + a_4^2 + a_5^2 + 2a_3a_4 + 2a_4a_5 + 2a_5a_3$$
$$\le 3\left(a_3^2 + a_4^2 + a_5^2\right) \le 3\left(\frac{1}{5} + \frac{1}{5} + \frac{1}{5}\right) = \frac{9}{5}.$$

Thus, $a_3 + a_4 + a_5 \le \sqrt{\frac{9}{5}} < \sqrt{2}$. This ensures that the five squares fit within the square of area 2 as shown in Figure 14.10. See much much more about packing squares in a square and other packing problems in *Further Explorations* E14 and E15. ∎

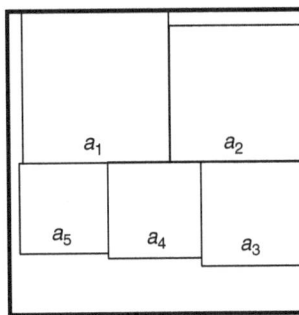

Fig. 14.10

Fifteenth Colorado Mathematical Olympiad
April 24, 1998

Historical Notes 15

A record four Olympians win First Prizes

Alexander Soifer (right) answering questions of John Crocket during the Olympiad Photo by Chuck Bigger, photographer of *The Gazette Telegraph*

Unlike a year ago, when the Colorado Mathematical Olympiad was buried by two feet of snow and had to be postponed, this year's event featured Colorado-blue skies and temperatures in the mid 70s. On April 24, 1998, the Fifteenth annual Colorado Mathematical Olympiad brought together some 500 middle and high school students. Contestants came from all over Colorado: Rangely, Durango, Denver, Thornton, Baker, Commerce City, Aurora, Lakewood,

A. Soifer, *The Colorado Mathematical Olympiad and Further Explorations:*
From the Mountains of Colorado to the Peaks of Mathematics,
DOI 10.1007/978-0-387-75472-7_29, © Alexander Soifer, 2011

Littleton, Castle Rock, Cascade, Sedalia, Parker, Fort Collins, Manitou Springs, Henderson, Lafayette, Longmont, Erie, Parker, Franktown, Falcon, Peyton, Calhan, Yoder, Ellicott, Simla, Cañon City, Fountain, Monument, Elbert, Black Forest, and Colorado Springs.

Thinking Heads. Photos by Chuck Bigger, photographer for *The Gazette Telegraph*

In an unprecedented show of talent, *four* contestants each solved four and a half problems out of five (only Olde Victorian Problem 5(b) survived all attempts). Each of the four musketeers was awarded first prize. They were Benjamin Coates, a senior student of Mrs. Terry Sloan from Manitou Springs High School; Aaron Parsons, a senior student of Melvin Oliver from Rangely High School; John Batchelder, a junior student of Dean Rockwell from Columbine High School; and John Crockett, a junior student of John Wicklund from Doherty High School.

Benjamin received a Compaq computer, while Aaron and the two Johns each received $1,000 scholarship to be used at any accredited American university of a four-year college. Each of the four winners received the gold medal of the Olympiad, the book *Colorado Mathematical Olympiad: The First Ten Years and Further Explorations* by

Alexander Soifer, and a Hewlett-Packard 48G calculator. Benjamin also received three other books by Alexander Soifer as a Creativity Award, for being the only contestant to observe a reduction of problem 3(b) to problem 3(a).

From the left, standing: Gold medalists John Crockett, Benjamin Coates, Aaron Parsons, and John Batchelder; sitting: Dean James A. Null, Alexander Soifer, and Isabelle S. Soifer (Isabelle passed Olympiad pins to all winners)

Second prize was awarded to Yogesh More, a junior from George Washington High School. He received the silver medal, a Hewlett-Packard 48G calculator with infrared printer, and the Olympiad book by A. Soifer.

Third prize was shared by three competitors: the youngest ever medalist Egon Cohen, a seventh grader (!) who was taking math at Cherry Creek High School; Joseph Alberto, a senior from Ellicott High School; and Tim Thiele, a senior from Fountain-Fort Carson High School. Each of them received the bronze medal, a Hewlett-Packard 48G calculator and the Olympiad book by A. Soifer. For such a high achievement at such a record young age Egon was also given a few "toys": memorabilia of the City of Colorado Springs, the University of Colorado, and the U.S. Space Command. While walking the rooms to answer kid's questions, I noticed that Egon Cohen was so eager to write his solutions that he did it standing up!

We also awarded 18 first honorable mentions and 65 second honorable mentions.

The Political Cartoon Award was presented to my son Mark S. Soifer, a junior from Palmer High School, for his graphic assessment of the political climate in Washington, D.C. Elisa Evans, a home schooled eighth grader from Calhan, received the Literary award for her two poems.

The Teacher Achievement Awards were presented to Doherty High School's John Wicklund, the teacher of a first prize winner in every of the past three years; and Rangely High School's Melvin Oliver, whose *eight out of eight* students won awards in this Olympiad, including a first prize, four first honorable mentions and three second honorable mentions. Rangely High School's Principal was in attendance. Note that Rangely is located in the Northwestern corner of Colorado, a 14-hour round trip by car to Colorado Springs!

Greg Hoffman (right), the most dedicated sponsor of the Olympiad

Appreciation Of The Judges was awarded to Roberta Wilson, President of the Palmer Alumni Association; Senator Jeffrey Welles, who spoke at the Awards Ceremony for the fifth time and was instrumental in funding education in the Colorado Senate for years; Councilman James Null, a supporter of the Olympiad for 10 years; and Greg Hoffman, for spearheading donations of computers in 1996 and 1997, and now the largest cash donation to date.

The prize fund of the Olympiad was generously donated by Schlage Lock Company, CASIO, GE Global IT Solutions, Texas Instruments, Hewlett-Packard, Colorado College, CU-Colorado Springs Bookstore, Colorado Springs School District 11, Air Academy School District 20, Harrison School District 2, Widefield School District 3, Rangely High School, Gateway High School, and the City of Colorado Springs.

The Award Presentation Program included *Review of Solutions of the Olympiad Problems* by Alexander Soifer and lecture *Mathematics of Map Coloring* by Alexander Soifer.

Alexander Soifer presenting the annual *Review of Solutions*. Victorian map coloring on the blackboard must have spilled on his sweater

The following guests of honor, hosts and sponsors addressed the winners and present the awards: Jeffrey M. Wells, Colorado State Senate Majority Leader (Senator Wells made it straight from the Senate session by 5:30 P.M.); Linda Bunnell-Shade, Chancellor; Elizabeth S. Grobsmith, Dean, College of Letters, Arts and Sciences; Paul Sale, Associate Dean of the School of Education, Douglas E. Swartzendruber, President, UCCS Faculty Senate – all from UCCS; James A. Null, Colorado Springs Councilman; Greg Hoffman, Personnel Manager, Schlage Lock Company; and Mark Postlewate, Public Sector Sales Manager, and Karen Golden, Public Sector Account Executive – both from GE Capital IT Solutions, Denver.

Organizers of Olympiads need to know that everything is seldom perfect. It was hilarious to see the great company Hewlett-Packard supplying engraved brass plates, advertising that the calculators were "donated by HP", while cutting their contribution to one-fifth of that previous years (and soon would cease contributing at all). My Chancellor Linda Bunnell-Shade refused my – and my dean's – request to continue a petty $995 donation to the Olympiad that she and her predecessors had provided annually for seven years. She will surprise me even more – see *Historical Notes 18*.

Nonetheless, the Olympiad compiled quite a record. In the 15 years of the Colorado Mathematical Olympiad, over 10,000 students have participated during 1984-1998. They have written some 50,000 essays, and were awarded some $100,000 in prizes.

Every year between 1996 and 2005, as a member of the USA Mathematical Olympiad (USAMO) Subcommittee, I attended the "Bestowing of Awards and Olympiad Address" at the National Academy of Sciences and "Reception and Dinner" at the Diplomatic Reception Room of the United States Department of State. One of the nine receptions stands out in my memory: a cabinet member, the President's Advisor on Science and Technology, walked straight toward me with his arms open for a hug. This was a person photographed in *Historical Notes* 2. I am talking about Neal Lane, who in 1985 was the chancellor of my campus of the University of Colorado and in 1998 President Clinton's advisor on science and technology. Someone took our photograph to commemorate the reunion.

With the President's Advisor on Science and Technology and former UCCS Chancellor Neal Lane (left) at the United States Department of State

Problems 15

Problems 15.1, 15.2, 15.3, and 15.4 were created by me especially for this year's Olympiad. Problem 15.5 is another matter. In my historical-mathematical research for *The Mathematical Coloring Book* [S11], I read a good number of nineteenth-century Victorian mathematical papers. Clearly, the precision and rigor of mathematical prose has improved since then, but something charming was lost – perhaps, we have lost the "taste of time" in our demand for "objective," impersonal writing, enforced by journal editors and many publishers. I decided to give a historical taste to my Olympians, and show them that behind Victorian clothing we can find the pumping heart of the Olympiad spirit.

Thus, I took problem 15.5(a) from the last page of the Victorian 1879 paper by Alfred Bray Kempe, in which the author published his famous attempted proof of the Four-Color Conjecture, 4CC (a mistake was found only 11 years later!). I got problem 5(b) in another Victorian paper, from 1880, by Peter Guthrie Tait, in which the author attempted to prove the Four-Color Conjecture as well (4CC was finally settled a century later, in 1976, by Kenneth Appel and Wolfgang Haken – see more in [S11]).

From time to time, I also like to poke fun at contemporary politics. This year, the Independent Counsel Kenneth W. Starr inspired (or should I coin a new word *dispired*, since I truly disliked this political nitpicking fellow) my creation of problems 15.3–15.4. You may recall, he was charged with investigating the suicide of the deputy White House counsel Vince Foster and the Whitewater land transactions of President Bill and Mrs. Clinton. His uncovering of the Monica Lewinsky affair prompted the President to lie about the affair and thus caused the U.S. Congress to bring forward articles of impeachment, not approved by the U.S. Senate. In fact, my Problem 15.4 title, "Investigation Continues," appeared a few weeks after I created the problem, on April 3, 1998, as a subtitle of the Associated Press article "Starr Unfazed by Dismissal of Jones Suit: Investigation Continues"!

The problems were selected and edited for you by the Problem Committee: Gary Miller and Alexander Soifer.

15.1 THE GREAT DIVIDE (*A. Soifer*).

All digits of a 1998-digit number A are 1. Find the greatest common divisor of A and 1111.

15.2 AMERICAN HIGHWAY (*A. Soifer*).

The United States has 1998 cities, some pairs of which are connected by highways. Every city is connected to at least 999 other cities. Prove that the highway system allows one to travel from any city to any other city with at most one change of highway.

15.3 WASHINGTON recTANGLES (*A. Soifer*).

(a) If Democrats and Republicans fill the 2×2 Lafayette Square, there will be two folks of the same party separated by a distance of at least $\sqrt{5}$.

(b) A 2×3 rectangular banner is colored in three colors (red, white, and blue, of course). Prove that there are two points of the same color separated by a distance of at least $\sqrt{5}$.

15.4 INVESTIGATION CONTINUES (*A. Soifer*).

(a) A 2×4 rectangular region R is colored in 4 colors. *Sources close to the investigation* inform us that the four corners of R are not colored two in one color and two in another color. Prove that there are two points of the same color at least a distance $\sqrt{5}$ apart.

(b) Show that the condition on the coloring of corners, leaked from the investigation, is essential: without this condition there is a 4-coloring of R with no distance $\sqrt{5}$ or greater between two points of the same color.

15.5 OLDE VICTORIAN MAP COLOURING (*A.B. Kempe*, 1879 and *P. G. Tait*, 1880).

Coloring of a map is an assignment of colors to each of the regions of the map (one color per region) such that no adjacent regions (i.e., regions that have a piece of boundary in common, not a mere finitely many points) get the same color. Let n be a positive integer; a map M is called *n-colorable* if there is a coloring of M in n colors.

(a) Prove that a map formed in the plane by finitely many circles is 2-colorable.

(b) A finite set S of points p, some pairs of which are connected by curves that do not intersect each other except at the points p, is called a *triple-delight* if each curve can be assigned one color, red, white or blue, so that each point in S is an endpoint of exactly one red, one white and one blue curve.

Prove that a map whose boundaries form a triple-delight is 4-colorable.

Solutions 15

15.1. Observe first that we can get a number with ten consecutive digits "1" as follows:

$$1111111111 = 1111 \times 1000100 + 11.$$

Similarly, since $1998 = 4 \times 499 + 2$, we can represent A as follows:

$$A = 1111 \times B + 11,$$

where $B = 100010001000...100$ is a number consisting of 499 digits "1" separated by three zeros, and followed by two zeros, i.e.,

$$A - 1111 \times B = 11.$$

But this means that the greatest common divisor $gcd\,(A, 1111)$ of A and 1111 must be a divisor of 11. It is easy to see that both A and 1111 are divisible by 11, and thus $gcd\,(A, 1111) = 11.$ ∎

15.2. Let A and B be two of the cities. If A and B are connected by a highway, we are done. Assume that A and B are not connected by a highway. There are 1996 other cities in the U.S., and A (as well as B) is connected by a highway to at least 999 of them. Since $999 \times 2 > 1996$, there is a city C among these 1996 cities such that both A and B are connected to C by highways (please see Figure 15.1). We are done. ∎

Fig. 15.1

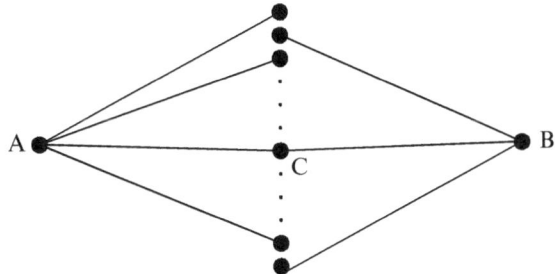

15.3. (a). Assume that the 2×2 square region S is colored in two colors in such a way that every pair of points of the same color is less than $\sqrt{5}$ apart. Draw a unit square grid on S(Figure 15.2).

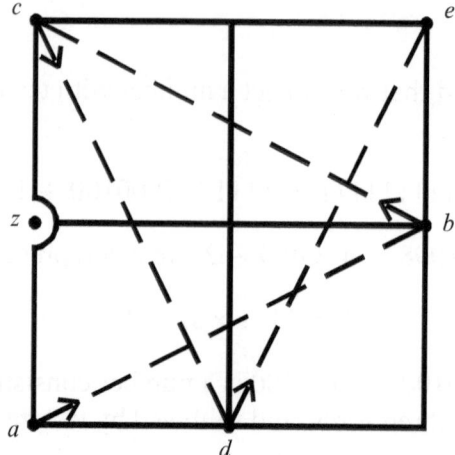

Fig. 15.2

Since the distance between every two consecutive points in a sequence a, b, c, d, e is $\sqrt{5}$, their colors alternate. Therefore, the points a and e have the same color, but they are farther than $\sqrt{5}$ apart, a contradiction. Therefore, in any coloring of S there are two points of the same color at a distance of at least $\sqrt{5}$.

Observe: we have not used the color of the point z in the solution! ■

15.3. (b). Assume that the 2×3 rectangular banner B is colored in 3 colors, red, white, and blue, in such a way that every pair of points of the same color is less than $\sqrt{5}$ apart. Draw a unit square grid on B (Figure 15.3).

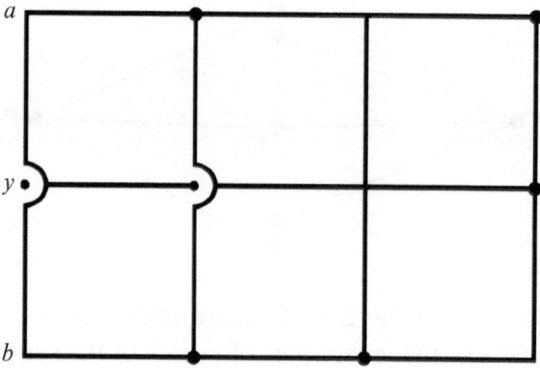

Fig. 15.3

Since we used three colors, and the banner B has four corners, at least two of the corners, a and b must be assigned the same color, say white. Points a and b must be endpoints of the same short side of B, for otherwise they would be farther than $\sqrt{5}$ apart. But then each bold point of the grid must be colored red or blue, for otherwise we would have a distance greater than $\sqrt{5}$ between a bold point and one of the corners a and b.

We now find ourselves in the setting of Problem 3(a), and its solution, applied to the grid of the bold points, carries over.

Observe: we have not used the color of the point y in the solution! ■

15.4. (a). Assume that the 2×4 rectangular region R is colored in four colors in such a way that every pair of points of the same color is less than $\sqrt{5}$ apart. Draw a unit square grid on R (Figure 15.4).

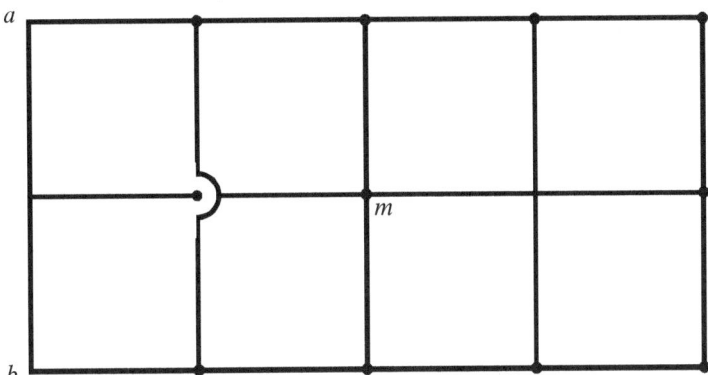

Fig. 15.4

Observe that the midpoint m of R is at a distance precisely $\sqrt{5}$ from every corner of R. Therefore, the color assigned to m may not be used in coloring the corners of R. Therefore, at most three colors were used in coloring the four corners of R, and thus there are two corners, a and b, of the same color! Points a and b must be endpoints of the same short side of R, for otherwise they would be farther than $\sqrt{5}$ apart. But then bold points of the grid cannot share a color with a and b, for otherwise we would get a distance at least $\sqrt{5}$ between a bold point and one of the corners a and b. Thus, all bold points must be colored in the remaining three colors.

We now find ourselves in the setting of Problem 3(b), and its solution, applied to the grid of the bold points, carries over, *or so it seems at first glance*. A closer analysis shows the need for one extra observation. In the solution of 3(b) we proved the existence of the two corners a' and b' of the same color in the 2×3 rectangular banner B. Points a' and b' must be endpoints of the same short side of B. It matters, however, which short side! Here the leak from the investigation comes in handy: the points a, b, a' and b' may not comprise the four corners of the rectangle R, for otherwise *the four corners of R would be colored two in one color and two in another color*. Therefore, a' and b' are in the column next to a and b (Figure 15.5), and the rest of the solution for 3(b) indeed carries over like a charm. ■

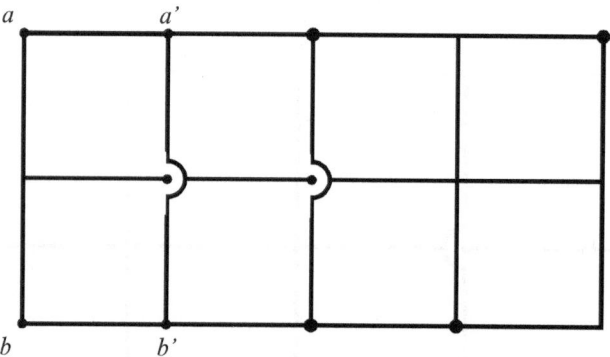

Fig. 15.5

15.4. (b). Let the four corners of the 2×4 rectangular region R be colored two in one color and two in another color. Our task is to expand this coloring to a 4-coloring of R with no distance $\sqrt{5}$ or greater between two points of the same color. Indeed, it can be done: see Figure 15.6 for such a 4-coloring of R, where the colors 1, 2, 3, 4 mark the respective regions. Vertices depicted empty (or white) are not included in the color of the small rectangle they appear in. ■

15.5. (a). Given a map formed in the plane by finitely many circles (Figure 15.7).

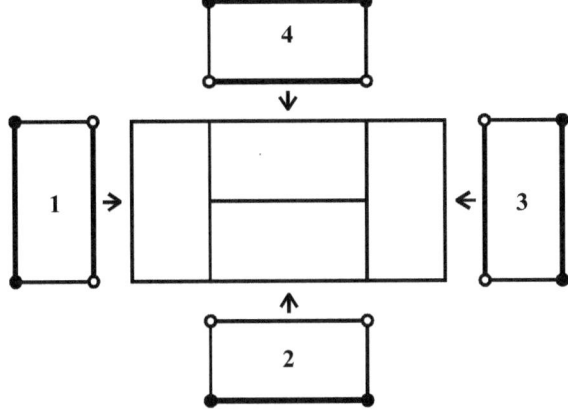

Fig. 15.6

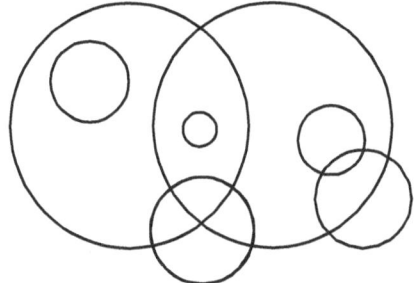

Fig. 15.7

We can partition all regions of the map into two classes (Figure 15.8): those contained in an even number of circles (color them black), and those contained in an odd number of circles (color them white).

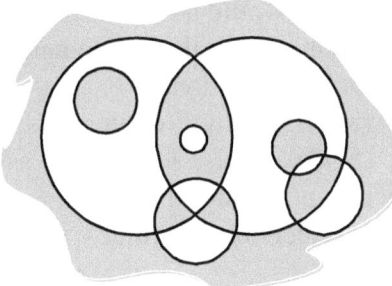

Fig. 15.8

Every time we cross a boundary between two regions of the map, the parity (of the region we are in) changes, and also the color of the region changes. Neighboring regions have thus been assigned different colors, and we are done.

Observe: the shape of a circle is of no consequence. We can replace circles in problem 15.5(a) by simple closed curves. This observation will be used in the solution of problem 15.5(b). ∎

15.5. (b). Let the boundaries of a map M form a triple-delight and be colored in three colors, red (r), white (w) and blue (b). Remove all red boundaries and look at the resulting map M_{wb}. Its boundaries are naturally partitioned into simple closed curves (because every vertex of M_{wb} has degree 2). Therefore, by Problem 15.5(a), the map M_{wb} can be 2-coloured, call the colors A and B.

Similarly, remove all white boundaries and look at the resulting map M_{br}. Its boundaries are naturally partitioned into simple closed curves. Therefore, by Problem 15.5(a), the map M_{br} can be 2-coloured, call them colors 1 and 2.

We have thus assigned every region of the map M one of the following *four* pairs of colors: $A1$, $B1$, $A2$ and $B2$. It is easy to verify that we have ended up with a proper 4-colouring of the map M. Indeed, assume that two regions, R_1 and R_2, that share a boundary b are assigned the same color, say $A1$ (due to symmetry of color assignments, we do not thus lose generality). What can possibly be the color of b?

It cannot be white or blue, for otherwise when we first removed only red boundaries (and kept white and blue ones!), we would not have assigned the same color A to R_1 and R_2, for in the map M_{wb} they were neighbors and had to be assigned different colors. It cannot be red, for otherwise when we next removed only white boundaries (and kept red and blue ones), we would not have assigned the same color 1 to R_1 and R_2, for in the map M_{br} they were neighbors and had to be assigned different colors. We are done!

Problem 15.5(a) was solved by 7 contestants; nobody solved the Victorian Problem 15.5(b). Thus, let us not assume that we are brighter than our Victorian ancestors!

Victorian Coloring Excursion continues in *Further Exploration* E16. ∎

Sixteenth Colorado Mathematical Olympiad

April 23, 1999

Historical Notes 16

Young mathematicians prevail in tragic circumstances

Tuesday April 20, 1999, was the gravest day in the history of Colorado. The tragic shooting at Columbine High School near Denver left many students and a teacher dead, others wounded. It shocked every one of us. I was torn by a moral dilemma: should I cancel the Colorado Mathematical Olympiad, scheduled for the coming Friday? Columbine often participated. Last year Columbine's junior John Batchelder won first prize. Was he OK? If not, I felt, then cancellation was a must. If he were OK, what would he prefer, a cancellation in solidarity with Columbine's tragedy or holding the Olympiad as a symbol that we were down but not out?

All evening the phone circuits were understandably busy. Late at night I got through. John was somber but physically unharmed. He said he would come on Friday to UCCS for the Olympiad. "OK, John, then we'll hold it," was my reply.

The following morning, on April 21, 1999, at 9:32 A.M., I sent an e-mail to all professors of my university asking them to join me and volunteer to teach at Columbine High School.

My e-mail was entitled "Teaching Aid for Columbine High":

The gravest day of Colorado ...

What can we do for Columbine High?

Teachers were shot...

A. Soifer, *The Colorado Mathematical Olympiad and Further Explorations:*
From the Mountains of Colorado to the Peaks of Mathematics,
DOI 10.1007/978-0-387-75472-7_30, © Alexander Soifer, 2011

We can form a list of those who wouldn't mind to come to Columbine and teach on some days until the end of this school year.

We will offer them a list of faculty in English, History, Philosophy, Foreign Languages, PE, Math, Physics, Biology, Chemistry, Geography, Computers, Econ., Art, Music, etc., etc. – and they will choose what they need.

Do let me know if you would volunteer.
If you are a department secretary, please, talk to faculty who do not use e-mail and let me know.
I will compile a list of teaching aid that UCCS can offer.

Today the school is closed.
I will talk later today to the district and John Batchelder and his father, and tomorrow to the principal ... if he is OK ...

    Yours,

    Alexander Soifer

A number of my fellow professors volunteered right away: Mike Ciletti of Electrical Engineering, Douglas Swartzendruber of Biology, Rick Wunderli of History, Kelli Klebe of Psychology, Terry Engel of English, Perrin Cunningham of Philosophy. I forwarded the list to the school district (the school was closed) right away, and to John Batchelder. He replied the same night of April 21, 1999, at 11:46 P.M.:

I am very grateful that there are people like you who show their concern to all those impacted by the tragic event at Columbine High School. I have received over one hundred e-mails from people throughout the United States, Canada, and even one from a high school student in Israel, all wishing to show their support to the victims of the incident. It is refreshing to know that there are people who not only feel sympathy towards the impacted individuals, but also put great effort into showing their concerns to the students and faculty of Columbine.
Sincerely,
John Batchelder
Columbine High School senior

Thursday April 22, 1999, brought a snowstorm throughout the State and made the cancellation of the Olympiad a possibility again. UCCS

Vice Chancellor O. Cleve McDaniel was in charge of the decision whether to close the campus. He and the University Police Chief Susan Szpyrka kept me informed. They did not make their decision by noon when Mel Oliver, a mathematics teacher at Rangely High School in the Northwestern corner of the State, had to make his decision whether to drive his students some 300+ miles to Colorado Springs. As updates were coming in, I was sitting glued to the computer screen:

Subject: Weather update.

Date: Thursday, 22 Apr 1999 16:24:51.

From: Susan Szpyrka, UCCS Police Operations.

To: O. Cleve McDaniel, Alexander Soifer.

Still impossible to predict what tomorrow will be like but here are the latest conditions. Due to the Math Olympiad, I am including areas around the state...

Dozens of e-mails from students, parents, and teachers popped up on the monitor. Here is one of them:

Dear Dr. Soifer, my son [...] is very interested in participating in the Math Olympiad competition, but we live in Ft. Collins and you know how the weather forecast is. Our school district has forbidden the school sponsored trip. [...] is asking me to drive him, and I am really hesitant because of the risks involved. Manfred wrote an e-mail to you earlier in the day, and you answered that you will postpone the competition only if the University is closed. But the University closes if there is danger for its students, and they don't come from Fort Collins, Denver, Pueblo, or Durango...

In fact Michelle Bohren and her mom had already left Durango, and Mel Oliver with his 10 students were already en route from Rangely to Colorado Springs. The decision had to wait till the wee hours of the morning. University Police had a long night ahead of them checking the forecast and the condition of the steep Austin Bluffs Parkway, where my university is located.

On Friday, April 23, 1999, at 5:30 A.M., I got a call from the University Police. The decision was made: the campus would stay open. The Olympiad was on. I got to my computer. E-mail messages to students, parents, and teachers in Denver, Littleton, Ft. Collins flew out like meteors: "We are on," "The Olympiad is on," "Start your engines!" I called those who left their phone numbers with me: "Parker Vista, Castle Rock, come when you can," "We will accommodate late arrivals." Even now in 2010, eleven years later, I feel the adrenaline flowing as I am writing these lines and thus reliving this memorable hectic day.

Some 300 students eventually braved the weather. They came from all over Colorado: Rangely, Durango, Littleton, Denver, Castle Rock, Parker, Manitou Springs, Franktown, Calhan, Ordway, Olney Springs, Peyton, Woodland Park, and Colorado Springs.

The April 22-23 snowstorm changed on Sunday, April 24th into hail and electric storm, then into rain, which went on for over a week, especially heavy rain fell on April 29 and the day of the Award Presentation, April 30. I was standing in a pool of water in my office, my feet soaking wet, as I was printing out the list of winners on my computer, raised from the floor by a piece of plywood. This was a trying year in the history of the Olympiad.

First prize was awarded to Yogesh More, a senior from George Washington High School in Denver. Last year's silver medalist, now Yogesh received the gold medal, a $2,000 scholarship to be used at any accredited American university or four-year college, a graphing calculator, and the book *Geometric Etudes in Combinatorial Mathematics* by Vladimir Boltyanski and Alexander Soifer. Yogesh was the only contestant to solve Problem 16.5.

Second prize was shared by two competitors: Charlie O'Keefe, a senior from Arapahoe High School in Littleton, who was the only contestant to advance in Problem 16.4a; and John Crockett, a senior from Doherty High School in Colorado Springs, a last year's winner. They both received the silver medal, a $1,000 scholarship to be used at any accredited university or four-year college, a graphing calculator, and the book *Geometric Etudes*.

Third prize was shared by five competitors: Lisa MacWilliams-Brooks, a senior from Doherty High School in Colorado Springs; Ken Brazier, a senior from Woodland Park High School; Kaloyan Kapralov, an exchange student from Bulgaria who at the time attended Fountain Valley School as a junior; Denny Smith, a junior from Palmer High School in Colorado Springs; and John Batchelder, a senior from Columbine High School in Littleton. Each of them received the bronze medal, a programmable calculator by Texas Instruments or Hewlett-Packard, and *Geometric Etudes*.

As you recall, John Batchelder was the gold medalist the previous year, and the one whose opinion I considered most important on whether to run the Olympiad this year. You may wonder what happened with John this year that he got "only" the bronze medal. A short answer is the Columbine massacre that happened three days before the Olympiad. But let me give you a longer answer. It was generous and courageous of John to come and thus allow me to run the Olympiad. However, there was nothing he could do with himself. He got up after two hours (out of four hours given him), submitted his Olympiad paper with perfect solutions of Problems 16.1, 16.2 and 16.3. "I cannot seat here when my friend is in the hospital in a serious condition," said John to me. I understood, of course, and asked him to work at home on the remaining hardest Problems 16.4 and 16.5. When a week later John Batchelder arrived before the Award Presentation, he handed in to me perfect solutions of 16.4 and 16.5!

We also awarded 4 fourth prizes, 20 first honorable mentions, and 50 second honorable mentions.

A "Nod of the Judges" award went to Shauna Uberecken, a sophomore from Coronado High School in Colorado Springs for her wit under pressure of competiton. Literary Award was presented to James Carroll, a sophomore from Arapahoe High School in Littleton (he will appear again in the next two *Historical Notes*):

Old Glory
By James Carroll

We marched into battle, our flag up above
Iy fluttered so freely, just like a dove.
We met the enemy, ten past nine,
They shot our ppor flag, times niineteen-nine-nie

We picked up the shreds.
Put them all in one place,
Bot a look of dspair was on everyone's face.
"Oh no!" Phil cried, "What shall we do?"
The biggest piece left, measures two by two!

James Carroll receiving the 1999 Literary Award. He will win third prize in 2000, and another Literary Award in 2001

Greg Hofman, Personnel Manager at Schlage Lock Company, had for several years been a major force behind sponsorship of organizations he belonged to: Apple Computer, SCI, Schlage Lock Company. We honored him with "Appreciation of the Judges" award.

The prize fund of the Olympiad was generously donated by Schlage Lock Company, CASIO, Colorado Springs School District 11, Air Academy School District 20, Harrison School District 2, Rangely School District RE-4, Douglas County School District RE-1, Texas Instruments, Hewlett-Packard, The Colorado College, CU-Colorado Springs Vice Chancellor for Academic Affairs John Pierce, CU-Colorado Springs Bookstore, John and Ardyce Putman and Alexander Soifer.

As you can see from the list, the sponsor from the Olympiad's inception in 1984 Hewlett-Packard disappeared from the list, and never came back.

The award presentation program included *Review of Solutions of the Olympiad Problems* by Alexander Soifer and lecture *Etudes on Maximum and Minimum* by Alexander Soifer.

The following guests of honor, hosts, and sponsors addressed the winners and presented the awards: Chancellor Linda Bunnell-Shade; and Greg Hoffman, personnel manager, Schlage Lock Company. I read greetings on behalf of Colorado Governor Bill Owens and Colorado Senate President Ray Powers.

Shortly after the Olympiad, the winner, Yogesh More, went to study at M.I.T. Silver medalist John Crockett entered the Utah State University, while another silver medalist, Charlie O'Keefe, went to the Colorado School of Mines.

Problems 16

Alexander Soifer contributed Problems 16.1, 16.2, 16.3, and 16.4. Problem 16.5 was inspired by the fine 1997 Russian book *How One Solves Non-Standard Problems* by Alexej Kanel-Belov and A.K. Kovaldzhi. The problems were selected and edited by the Problem Committee: Robert Ewell, Gary Miller, and Alexander Soifer.

16.1 OLD GLORY (*A. Soifer*). An 80×100 flag was shot in battle 1999 times. Prove that there is a 2×2 bullet-hole-free square in it.

16.2 COLORING NODES (*A. Soifer*). What is the smallest number of colors required for coloring nodes of a 1999×1999 square grid so that any segment connecting two nodes of the same color has a node of another color between them?

16.3 COLORING CELLS (*A. Soifer*).

(a) Is it possible to color some cells of an infinite square grid so that no matter how a 1999×1999 square is placed with all its sides on the lines of the grid, it contains exactly one colored cell?

(b) Would the answer change if we replaced the 1999×1999 square with a 1999×2000 rectangle?

16.4 EQUILATERAL POLYGONIA (*A. Soifer*). Is there an *equilateral* 1999-gon (all sides of equal length) with all vertices in nodes of a square grid? Is there such a 2000-gon? The polygons do not have to be convex and their sides may intersect.

16.5 REGULAR POLYGONIA (*A. Kanel-Belov and A.K. Kovaldzhi*). Is there a *regular* 1999-gon with all vertices in nodes of a square grid? Is there such a 2000-gon?

Solutions 16

16.1. Partition the flag by horizontal and vertical lines into 2×2 squares (Figure 16.1).

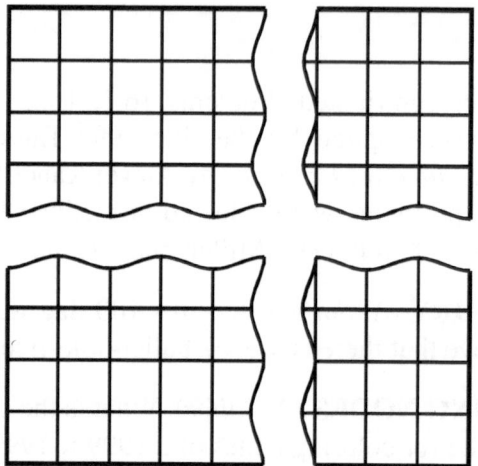

Fig. 16.1

There are $40 \times 50 = 2000$ squares, and only 1999 bullet holes, therefore, there is at least one square that is bullet-hole-free. ∎

16.2. Consider a unit cell of the grid. It has four nodes in its corners, all of which must be colored differently to satisfy the condition on coloring. By translating the colored four nodes we can achieve the 4-coloring of all nodes (see Figure 16.2).

This 4-coloring satisfies the given condition. Indeed, let A and B be two nodes of the same color, say color 1. Place a Cartesian coordinate system on the grid with the origin at A and axes parallel to the lines of the grid. Observe that a node is colored in color 1 if and only if both its coordinates are even integers. Then the coordinates m, n of B must

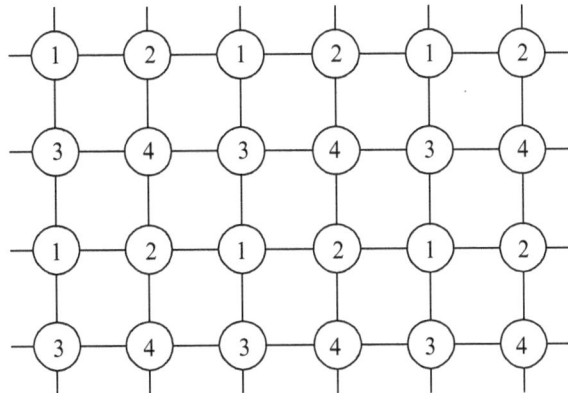

Fig. 16.2

be both even. Let 2^k be the highest power of 2 that divides both m and n, $k > 1$. Then the point $\left(\frac{m}{2^k}, \frac{n}{2^k}\right)$ is on the segment AB, between A and B, and is colored not 1 because at least one of its coordinates is not even. ∎

16.3.

(a) Take a 1999×1999 square and color its lower left unit square. Now tile the plane with translates of this 1999×1999 square. This delivers a required coloring.

(b) Assume that such a coloring exists. Let A be a colored unit square. Cover it by a 1999×2000 rectangle B_1 so that A lies in the lower left corner of B_1 (Figure 16.3). Observe that there must be no other colored unit squares inside B_1. Now cover A by another 1999×2000 rectangle B_2 so that A lies in the upper left corner of B_2. Observe that there must be no other colored unit squares inside B_2. But the union of B_1 and B_2 contains a 2000×1999 rectangle that does not cover A, and thus contains no colored squares, a contradiction. Therefore, such a coloring does not exist. ∎

16.4.

(a) Assume that an equilateral 1999-gon P with all vertices in nodes of a square grid exists and all its sides are of length w. Place a Cartesian coordinate system on the grid with the origin at A and

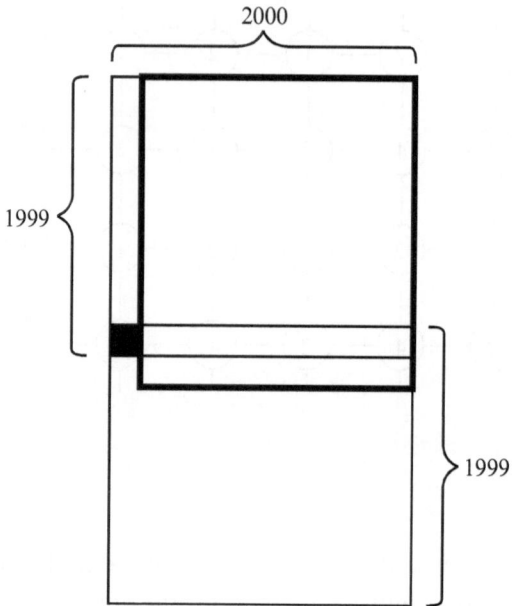

Fig. 16.3

axes parallel to the lines of the grid (Figure 16.4). Let the projections of the sides of P on the axes be respectively $x_1, x_2, \ldots, x_{1999}$
and $y_1, y_2, \ldots, y_{1999}$ (Figure 16.5). Then we get the following
equalities:

$$x_1 + x_2 + \ldots + x_{1999} = 0 \qquad (*)$$

$$y_1 + y_2 + \ldots + y_{1999} = 0 \qquad (**)$$

$$x_i^2 + y_i^2 = w^2 \text{ forevery } i = 1, 2, \ldots, 1999 \qquad (***)$$

We can assume without loss of generality that not all x_i, y_i are
even, for otherwise we can divide them all by the highest common
power of 2 and get a smaller equilateral 1999-gon with all vertices
in nodes of the grid. So, let for definiteness x_1 be odd. Then, of
course, $x_1^2 \equiv 1 \pmod 4$. We will consider two cases.

If y_1 is odd, then we get $w^2 = x_i^2 + y_i^2 \equiv 2 \pmod 4$, and the
equalities (***) imply that *all* x_i, y_i are odd, which contradicts the
equality (*): a sum of 1999 odd numbers cannot be even.

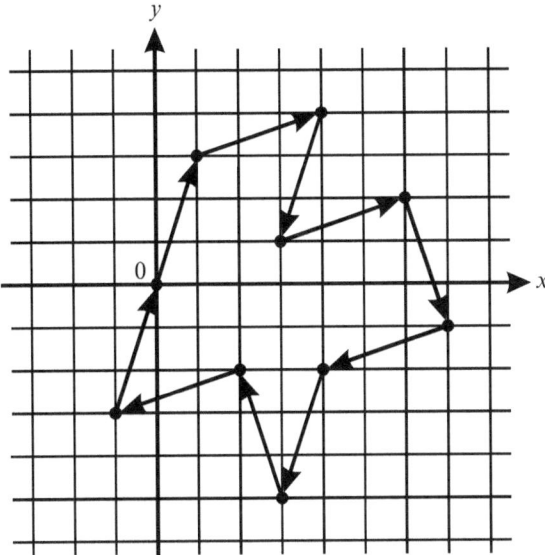

Fig. 16.4

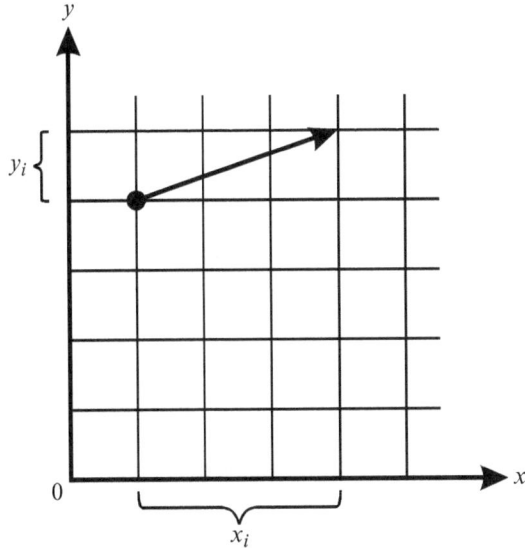

Fig. 16.5

If y_1 is even, then $w^2 = x_i^2 + y_i^2 \equiv 1$, (mod4) and the equalities (***) imply that in *every* pair x_i, y_i one number is odd and the other even, thus the total number of odd x_i, y_i is 1999. On the other hand, the equality (*) implies that the number m of odd x_i is even; and the equality (**) implies that the number n of odd y_i is even. Thus, we get $m + n = 1999$, a contradiction since the sum of two even numbers cannot be odd. ∎

(b) Behold (Figure 16.6):

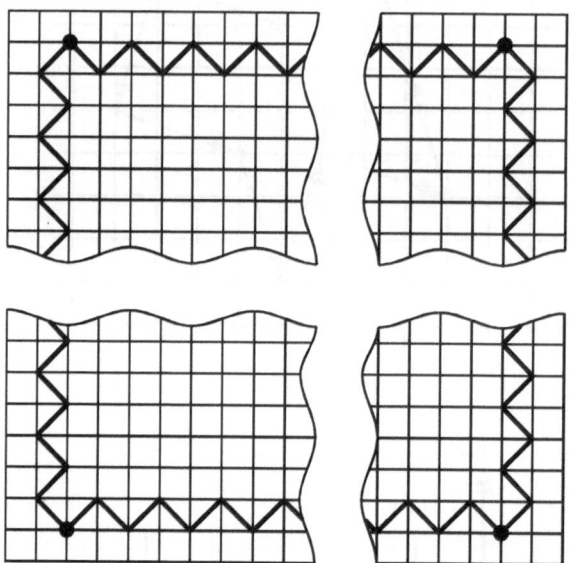

Fig. 16.6

Second Solution of Problem 16.4(a). The idea of this solution was found during the competition independently by the Arapahoe High senior Charlie O'Keefe and the judge of the Olympiad Ed Pegg, Jr. I provided a number-theoretic interpretation of the chessboard coloring and the missing proof of the impossibility of having non-monochromatic sides of the 1999-gon.

Assume that an equilateral 1999-gon P with all vertices in nodes of a square grid exists and all its sides are of length w. Color the nodes in a chessboard fashion black and white. We will prove that all 1999 vertices of P must lie in nodes of the same color.

Assume that it is not so. Then we can find two adjacent vertices A and B of P such that one, say A, is black and the other, B, is white.

Place a Cartesian coordinate system on the grid with the origin at a black node and axes parallel to the lines of the grid

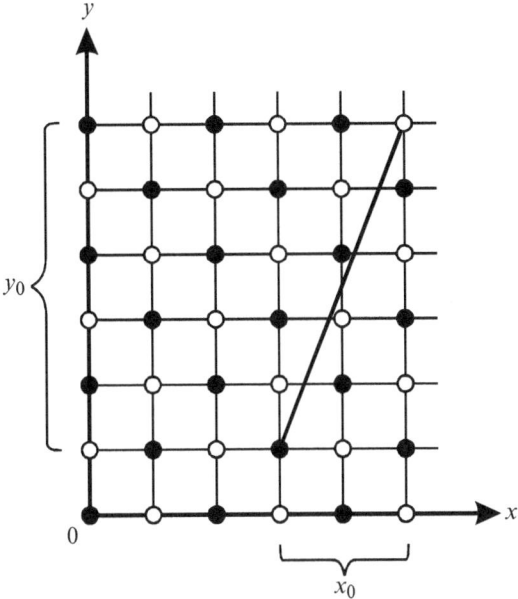

Fig. 16.7

(Figure 16.7). Then we can give a formal description of our chess-board coloring: black nodes are those whose coordinates are integers of the same parity (i.e., both even or both odd), while white nodes are those whose coordinates are integers of the opposite parity (i.e., one even and one odd).

We get

$$w^2 = x_0^2 + y_0^2,$$

where x_0 and y_0 are projections of AB on the axes respectively. But in view of the description of chessboard coloring above, the numbers x_0 and y_0 have opposite parity and thus w^2 is an odd integer. But this means that no segment of length w can possibly connect two nodes of the same color (for this would have made w^2 even!). And thus colors alternate as we walk along the perimeter of P, which implies that P has an even number of sides, and thus contradicts P being a 1999-gon.

Thus, we have proved that all vertices of P lie in black nodes. Erase all white nodes – we are left with a square grid (of black nodes) of unit size $\sqrt{2}$. Now color the new grid in two colors. Once again, all vertices of P will lie in points of the same color, say black. Erase all white nodes, etc. This process that does not affect the 1999-gon P, will produce grids of larger and larger unit size, yet with all vertices of P still in the nodes of all these grids. But eventually the unit size of the grid will become greater than, say, diameter of P, and then P will have a "hard time" having all its vertices in nodes of the grid. This contradiction proves that the required 1999-gon does not exist. ∎

16.5. *Remark.* Let a vector **a** start and end in nodes of a grid. If we translate the vector so that its starting point moves to a new node of the grid, then its ending point will get into a grid point as well.

Solution. Let $n \geq 7$. Assume that there is a regular n-gon P with all vertices in nodes of a square grid. Mark the direction on every side clockwise, thus making a vector out of every side (Figure 16.8). Now we can translate all starting points of side-vectors of P into one node of the grid. In view of the remark above, the ending points of sides-vectors will fall on the nodes of the grid and, due to symmetry, will form a regular n-gon P_1 that is smaller than P.

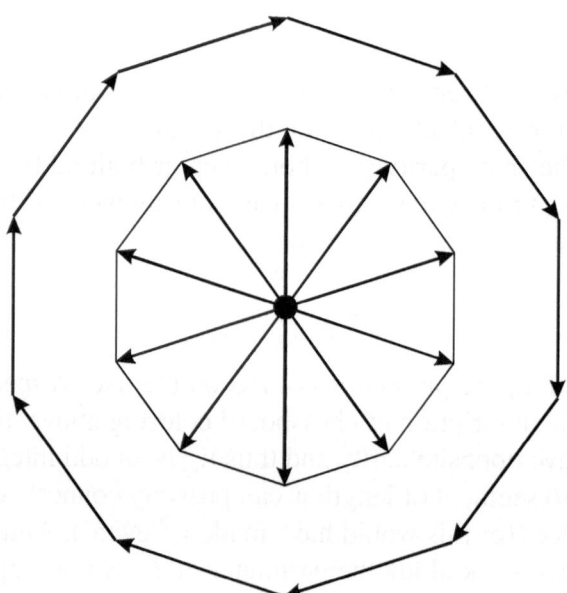

Fig. 16.8

Continuing this process (the infinite descent method, invented and loved by the great Pierre de Fermat), we will get a sequence of n-gons

$P, P_1, P_2, ..., P_n, ...$ whose vertices are in the nodes of the grid and whose side lengths approach zero. These two properties, of course, contradict each other. Thus, such an n-gon does not exist. ■

Second Solution of Problem 16.5 was found by Yogesh More, the gold medal winner, during the competition.

Let $n \geq 7$ or $n = 5$. Mark vertices of the polygon P with symbols $A_1, A_2, ..., A_n$ and translate each vertex A_i through the vector $A_{i+1}A_{i+2}$ (see Figure 16.9). The new positions $A_1', A_2', ..., A_n'$ of the vertices of P themselves form a regular n-gon P_1 with vertices in the nodes of the grid, which is inside of P and thus is smaller than P.

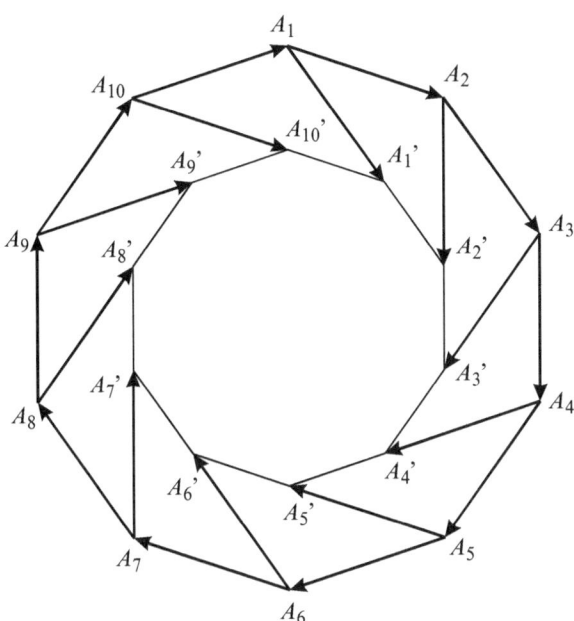

Fig. 16.9

Repeatedly applying this operation to P_1, we get an n-gon P_2, then P_3, etc. arriving at a contradiction. ■

Fig. 16.9

Seventeenth Colorado Mathematical Olympiad
April 21, 2000

Historical Notes 17

Fort Collins senior is finally golden

Clouded by the first anniversary of Columbine's tragedy, the seventeenth annual Colorado Mathematical Olympiad took place on April 21, 2000, under majestically blue Colorado skies and summer temperatures in the 70s. The last in the XX century Olympiad brought together 480 middle and high school students. Contestants came from all over Colorado: Rangely, Florissant, Commerce City, Aurora, Littleton, Adam City, Parker, Fort Collins, Manitou Springs, Henderson, Erie, Parker, Franktown, Peyton, Calhan, Hayden, Kiowa, Ellicott, Cañon City, Widefield, Monument, Elbert, Elizabeth, Larkspur, and Colorado Springs.

First Prize was awarded to Manfred Georg, a senior student of Kevin Follett from Ft. Collins High School. Manfred received the gold medal of the Olympiad, a graphing calculator, the first version of this book, *Colorado Mathematical Olympiad: The First Ten Years and Further Explorations*, and a $2,000 scholarship to be used at any accredited American university or four-year college. Manfred was one of only two contestants to solve the most challenging Problem 17.5.

Second prize was awarded to three contestants: Scott Danford, a junior from Ponderosa High School, Parker, Colorado; Mark Pond, a sophomore from Smokey Hill High School, Aurora, Colorado; and Will Horn, a junior from Arapahoe High School, Littleton, Colorado. Each of them received the silver medal of the Olympiad, top graphing calculator, and the Olympiad book by A. Soifer. In addition,

A. Soifer, *The Colorado Mathematical Olympiad and Further Explorations: From the Mountains of Colorado to the Peaks of Mathematics*, DOI 10.1007/978-0-387-75472-7_31, © Alexander Soifer, 2011

Scott Danford received Creativity Award: he was *the only one* of 480 contestants to solve Problem 17.3(b). This Award included two books on Geometry: *Geometric Etudes in Combinatorial Mathematics* by V. Boltyanski and A. Soifer and *How Does One Cut a Triangle?* by A. Soifer.

From the left: Silver medalist Scott Danford, Alexander Soifer, and Gold medalist Manfred Georg

Third prizes were presented to five competitors: Chris Barth, a sophomore from Arapaho High School; James Allred, a senior from Rangely High School; Sophia Knight, a sophomore from Littleton High School; Tim McCoy from Poudre High School, Ft. Collins; and James Carroll, a senior from Arapahoe High School. Each of them received the bronze medal of the Olympiad, a CASIO graphing calculator and the Olympiad book by A. Soifer.

The judges decided to award seven fourth prizes this year. One of fourth prize winners, Scott Greenwald, also received Creativity Award: he was one of only two of 480 contestants (the other was the first-prize winner) to solve Problem 17.5. We also awarded 22 first honorable mentions and 74 second honorable mentions.

The winner of a first honorable mention Bryce Herdt, also received Young Mathlete Award: this 7th grader achieved the highest result among all middle school contestants. Do not forget his name – he will shine in the next year's Olympiad!

The Prize Fund of the Olympiad had been generously donated by International Schlage Lock Company, CASIO, Colorado Springs School District 11, Air Academy School District 20, Harrison School District 2, Falcon School District 49, Hayden School District RE-1, Texas Instruments, Colorado College, CU-Colorado Springs Vice-Chancellor's Office, CU-Colorado Springs Bookstore, and the City of Colorado Springs.

It was disappointing and surprising to see Hewlett-Packard – and its offspring Agilent – abruptly dump sponsorship after 16 years of support, and in case of Agilent, even denying us a courtesy of reply. Industry is prone to short-term profit orientation, and only some enlightened industry leaders remedy it with a long term view and understanding of the importance of education. After all, they support education of their future employees and customers!

The Award Presentation Program included *Review of Solutions of the Olympiad Problems* by Alexander Soifer and his lecture *Y2K of Mathematical Education.*

The following guests of honor, hosts, and sponsors addressed the winners and presented the awards: Linda Bunnell-Shade, Chancellor; Elizabeth S. Grobsmith, Dean, College of Letters, Arts and Sciences; Paul Sale, Interim Dean of the School of Education, all from UCCS; Marci Morrison, Colorado State Congresswoman; Richard Skorman, Colorado Springs Councilman; and Greg Hoffman, Personnel Manager, Schlage Lock Company.

In the 17 years of the Colorado Mathematical Olympiad, some 12,000 students had participated during 1984-2000. They had written some 57,500 essays, and had been awarded over $120,000 in prizes.

I spent in Australia three weeks of May 2000 as a foreign advisor for Australian Mathematics Competitions. I enjoyed working with many talented colleagues and seeing my great old friends Peter Taylor, Warren Atkins and Graham Pollard. It was a thrill to visit again with the legendary George Szekeres, Esther Klein-Szekeres and Bernhard Neumann: I jotted down what only they would know about their young years in Budapest of the early 1930s, and these lines entered *The Mathematical Coloring Book* [S11].

Problems 17

Problems 17.1 and 17.5 are A. Soifer's variations on problems from Russian books *Instructive Games* by Sergei L. Tabachnikov and Andrei L. Toom (1987), and *Selected Problems and Theorems of Elementary Mathematics* by David O. Shklyarsky, Nikolai N. Chenzov, and Isaak M. Yaglom (1950) respectively. Jean Turgeon's research inspired me to create Problem 17.2 for which I found a new short solution. In 1976 Frank Harary invented a host of new tic-tac-toe style games; I "dressed up" one of them in a story to produce Problem 17.3. I created Problem 17.4 "from scratch" for this CMO-2000.

These problems were selected and edited by the Problem Committee: Robert Ewell, Gary Miller, and Alexander Soifer.

17.1 THE GOOD, THE BAD, AND THE UGLY (*A. Soifer*). Having taken care of the Ugly, the Good and the Bad are dividing a bounty of 2000 silver dollars as follows. First, the Good divides the coins into two piles with at least two coins in each pile. Then the Bad divides each of those two piles into two (nonempty) piles and takes the largest and the smallest of the four piles, leaving the other two piles for the Good. What is the maximum share the Good can count on no matter how smart – or greedy – the Bad is?

17.2 STONE-AGE ENTERTAINMENT (*A. Soifer and J. Turgeon*).

(A) Fred Flintstone and Barney Rubble in turn take pebbles from a pile of 2000 pebbles. They can take 1, 7, or 13 pebbles at a time. The player who takes the last pebble wins. Find a strategy that allows Fred or Barney to win regardless of how the other one may play. Oh, Fred goes first.

(B) What is the winning strategy if the players are allowed to take 1, 2, 7, or 13 pebbles at a time?

17.3 MORE STONE-AGE ENTERTAINMENT (*F. Harary and A. Soifer*).

(A) Fred Flintstone and Barney Rubble in turn color unit squares of a 2000 × 2000 square grid, one unit square per move. Fred uses red, and Barney uses blue. Fred wins if he gets a red 5-unit-square cross (that is, a unit square with its adjacent squares above, below, right, and left). Otherwise Barney wins. Find a strategy that allows Fred or Barney to win regardless of how the other one may play. Fred goes first.

(B) What is the winning strategy if the 5-unit-square cross is replaced by a 2×2 square?

17.4 TAKE A FLY (*A. Soifer*).

(A) 2000 pigeons sit in 2000 pigeonholes, one per hole. They all fly off and land in the same 2000 holes, one per hole, but no pigeon returns to the same hole. Define a *Nomad Camp* as a group of pigeons such that none of them landed in a pigeonhole previously occupied by another member of the same Camp. Prove that in any landing there exists a Nomad Camp of at least 667 pigeons.

(B) Show that there is a landing that does not contain a Nomad Camp of 668 pigeons.

17.5 MILLENNIA SQUARED (*A. Soifer*).

Given a square 2000×2000 table A of positive real numbers; let a_{ij} denote the entry that appears in row i and column j. Suppose that the equality $a_{ij}a_{jk}a_{ki} = 1$ holds for any i, j, and k between 1 and 2000. Prove that there are 2000 numbers $c_1, c_2, \ldots, c_{2000}$ such that $a_{ij} = c_i / c_j$ for any i and j between 1 and 2000.

Solutions 17

17.1. In order to maximize his share, the Good must divide the bounty into equal 1,000-coin piles. Then no matter how the Bad subdivides the two 1,000-piles, the largest and the smallest of the 4-piles will come from the same 1,000-pile (for the complement of the largest pile is the smallest one!), and thus add up to 1,000.

If the Good divides the 2000 silver dollars into unequal piles that contain m and n coins respectively, $m > n$, then the Bad can subdivide the m-pile into $m - 1$ and 1, and thus the largest pile will be $m - 1 \geq 1,000$.

After adding the smallest pile the Bad would positively have more than 1,000 coins. ■

17.2(A). Since 2000 is even and 1, 7, and 13 are odd, every one of Fred's moves would leave an odd number of pebbles in the pile. The winner, after the winning move, will leave 0 pebbles, and 0 is not an odd number. Therefore, Fred cannot win. Since somebody will win (as long as the game is not over, a move of taking 1 pebble is always available, and with every move the number of pebbles in the pile goes down), it must be Barney. Barney wins no matter how he plays.

17.2(B). Observe that 1, 7, and 13 all give remainder 1 upon division by 3; while 2 and 2000 give remainder 2 upon division by 3. Now the strategy is clear: Fred takes 2 pebbles on his first move, leaving 1998, a number divisible by 3. Then, if Barney takes 1, 7 or 13 pebbles, Fred takes 2; if Barney takes 2 pebbles, Fred takes 1, 7, or 13. In either case, Fred will leave after his move a number of pebbles divisible by 3, eventually leaving 0 pebbles (0 is divisible by 3 :-), and thus winning. ∎

17.3(A). Barney tiles the given 2000 × 2000 square grid by dominoes and comes to play with this home-prepared template (Figure 17.1). Every time Fred colors one square of a domino red, Barney colors the second square of the same domino blue. This guarantees that there are no completely red dominoes in Barney's tiling. On the other hand, no matter where you placed the 5-unit-square cross on the grid, it would cover completely at least one domino in Barney's tiling. Thus, Barney wins. ∎

17.3(B). Observe that the same strategy would not work, because it is possible to place a 2 × 2 square on the grid so that it does not cover completely any domino in Barney's tiling. However, we can change the template a little to make it work: just translate consecutive rows of the template through the width of one unit square (Figure 17.2), and let Barney use this template in following the same strategy of coloring the second square of the domino. In this template we have some monominoes (single unit squares) along the boundary. If Fred colors one of the monominoes, let Barney color another monomino. ∎

Second Solution of Problem 17.3(B). Something amazing happened with me during the week between the Olympiad and the Award Presentation Ceremonies. I found a new solution of this problem on...

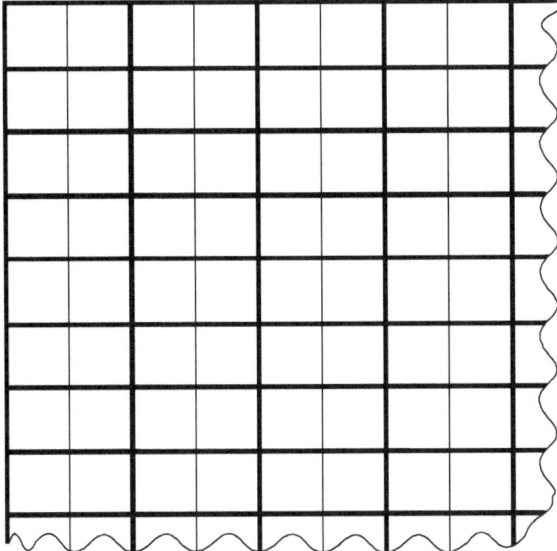

Fig. 17.1

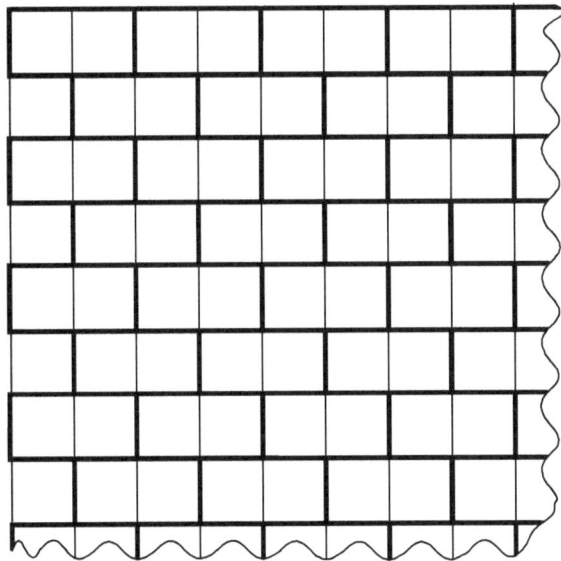

Fig. 17.2

an ancient 6000-year old seal! You can see it in Figure 17.3. The Tuesday, April 25, 2000, auction described it as follows:

Fig. 17.3

Ancient Persian Seal Stamp Sialk 4000 BC

A very rare and large Chalcolithic period[1], of Shalk II/III, Persian pottery impressed seal, or seal stamp. Probably used as both an official pass, and a design impression marker. A rare type found at Sialk, Rey, Susu and, or Tepe Giyan, generally north of Tehran

---

[1] The *Chalcolithic* is the name given to the period in the Near East and Europe after the Neolithic period and before the Bronze Age, between approximately 4500 and 3500 BC.

or near Kashan, and dating from 4200-3400 B.C. Very fine, large and in perfect condition, with buff orange body, deep design, and untouched burial surface. A rather esoteric museum type item, or for a serious collector of the formative period of seals. 4″ long x 2 1/4″ wide x 1 1/16″ deep.

As you can see, the ancient seal presents a "zigzag" tiling of the plane with dominoes. This tiling, taken as a template (please see Figure 17.4), allows us to use exactly the same strategy as we used in the first solution of this problem:

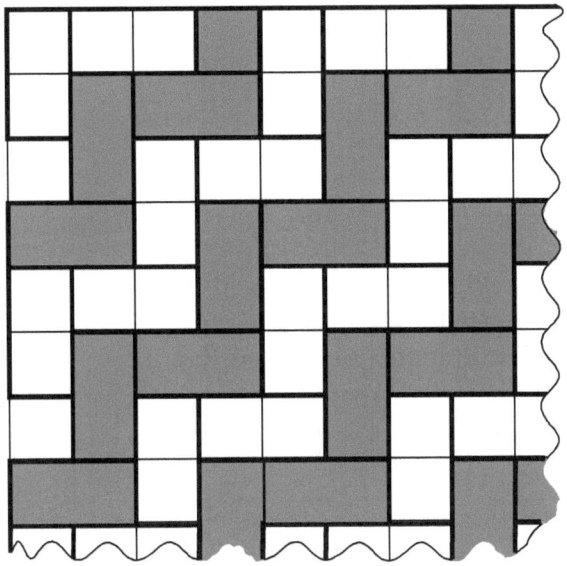

Fig. 17.4

In order to win, Barney uses this template and colors the second square of the domino the first square of which was colored by Fred. If Fred colors one of monominoes, Barney colors another monomino.

You may be wondering what happened to the ancient seal at the auction. I had to buy it as a memento of such an unusual way of discovering a mathematical solution!

See more about achievement games in *Further Exploration* E17 in this book. ∎

17.4(A). Number the pigeons with integers 1 through 2000. Pick any pigeon, say, the pigeon number a_1. Let it land in the pigeonhole occupied originally by the pigeon a_2. Let pigeon a_2 land in the pigeonhole originally occupied by pigeon a_3, and so on. This process of building a sequence of distinct pigeons cannot go forever because we have finitely many pigeons. But why would it stop? It would stop because we would get the same pigeon for the second time! This creates a *cycle*; we will use the notation $(a_1, a_2, \ldots, a_n)$ to indicate that pigeon a_i landed in the pigeonhole originally occupied by pigeon a_{i+1} for $i = 1, 2, \ldots, n - 1$, and pigeon a_n landed in the pigeonhole originally occupied by pigeon a_1. If this cycle does not exhaust all 2000 pigeons, we pick a pigeon b_1 outside of the cycle $(a_1, a_2 \ldots, a_n)$, and proceed in the fashion, thus creating a cycle $(b_1, b_2, \ldots, b_m)$. By continuing this process, we partition the 2000 pigeons into disjoint cycles:

$$(a_1, a_2, \ldots, a_n)(b_1, b_2, \ldots, b_m) \ldots (h_1, h_2, \ldots, h_t).$$

We can now select all even numbered pigeons from each cycle and combine them into a set C:

$$C = \{a_2, a_4, \ldots, b_2, b_4, \ldots, h_2, h_4, \ldots\}$$

The set C is a nomad camp. Indeed, all pigeons from C landed in odd-numbered pigeonholes.

Let us now assess number of pigeons $|C|$ in C:

$$|C| = \left\lfloor \frac{n}{2} \right\rfloor + \left\lfloor \frac{m}{2} \right\rfloor + \ldots + \left\lfloor \frac{t}{2} \right\rfloor,$$

where $n, m, \ldots, t$ are the lengths of our cycles, and $\lfloor a \rfloor$ stands for the largest integer not exceeding a. But for even n we have $\left\lfloor \frac{n}{2} \right\rfloor = \frac{n}{2}$. For odd n we get $\left\lfloor \frac{n}{2} \right\rfloor = \frac{n}{3}$ when $n = 3$, and $\left\lfloor \frac{n}{2} \right\rfloor > \frac{n}{3}$ when $n > 3$. Observe also, that all our cycles cannot have length 3 simply because 2000 is not divisible by 3. We are done, because

$$|C| = \left\lfloor \frac{n}{2} \right\rfloor + \left\lfloor \frac{m}{2} \right\rfloor + \ldots + \left\lfloor \frac{t}{2} \right\rfloor > \frac{n}{3} + \frac{m}{3} + \ldots + \frac{t}{3} = \frac{2000}{3}$$

and thus $|C| \geq 667$. ∎

17.4(B). Consider a landing that decomposes the 2000 pigeons into 667 cycles: 666 cycles of length 3 and 1 cycle of length 2.

Assume there is a nomad camp of 668 pigeons. Since we have 668 pigeons coming from 667 cycles, by the Pigeonhole Principle (no pun intended :-) we conclude that there will be two pigeons that come from the same 3-cycle or 2-cycle. But then one of these two pigeons would land in the pigeonhole originally occupied by the other pigeon, which contradicts our assumption of having here a nomad camp. Thus, there are no nomad camps of 668 pigeons in this landing. ∎

17.5. Let $i = j = k$. Then the given equality becomes $(a_{ii})^3 = 1$, and thus, $a_{ii} = 1$. Let now $j = k$. Then the given equality becomes $a_{ij}a_{jj}a_{ji} = 1$, and since $a_{jj} = 1$, we get $a_{ij} = \frac{1}{a_{ji}}$.

Now we can rewrite the given equality as follows:

$$a_{ij} = \frac{a_{ik}}{a_{jk}}.$$

Since the left side of the last equality has no k, the ratios on the right side are equal to each other for all values of k. This means that the rows are proportional. This also means that we can set $k = 1$ and choose the required c_i in a very simple way: $c_i = a_{i1}$ for all i between 1 and 2000. ∎

Eighteenth Colorado Mathematical Olympiad

April 20, 2001

Historical Notes 18

At the start of the new millennium, the Olympiad gets
the youngest winner ever: a Colorado Springs eighth-grader!

The Olympiad season commenced on April 19, 2001, with Raquel Rutledge's article "Let the math games begin" in the *Gazette*:

Solving one of Alexander Soifer's math problems is like finding a treasure.

"It feels like you found the largest-ever mushroom in the mountains or if you found an ancient Greek coin of Alexander the Great," said Soifer, a math professor at the University of Colorado at Colorado Springs.

Hundreds of students from around the state will join in the four-hour search Friday to solve five of Soifer's famous math problems.

It's the 18th annual Mathematical Olympiad sponsored by CU-Springs, and nobody has ever solved all five.

Even so, trying is fun, said two-time winner Matt Kahle.

Kahle, now preparing for a Ph.D. in mathematics at the University of Washington, won the contest in 1990 and again in 1991 when he was an Air Academy High School student.

"It was always something that was just fun," Kahle said. "The problems were interesting. The solutions were pretty. I liked the aesthetics of it."

A. Soifer, *The Colorado Mathematical Olympiad and Further Explorations: From the Mountains of Colorado to the Peaks of Mathematics,* DOI 10.1007/978-0-387-75472-7_32, © Alexander Soifer, 2011

Kahle, who will return to Colorado Springs to help judge this year's Olympiad, has some advice for contestants: Take your time. Use all four hours and doodle on scratch paper. And expect to be exhausted when you're finished.

"I was so mentally exhausted I couldn't even remember my own phone number," Kahle said.

The Olympiad is open to high school and middle school students.

About $6,000 in scholarships and other prizes such as scientific calculators, computer software and T-shirts will be awarded...

Soifer said many of his problems stem from old Russian mathematical folklore and are [meant] to elicit creativity.

Winners must be able to solve problems unlike anything they have ever seen before.

"They are original thinkers," he said. "They are all, without exception, curious people. They are creative. Each has a spark of brilliance."

Soifer said past winners, like Kahle, have often gone to earn Ph.D.'s. But not all were straight-A students in high school, he said. They just needed something to inspire them.

"My goal is to excite students, show them mathematics can be beautiful, surprising, elegant," he said.

The only hint Soifer gives to finding the answers:

"Solutions are short, it's just very hard to find them. It's like a treasure hunt."

The eighteenth Annual Colorado Mathematical Olympiad took place on April 20, 2001, the second anniversary of the Columbine tragedy. The first Olympiad in the twenty-first century brought together 480 middle and high school students. Contestants came from all over Colorado: Aurora, Calhan, Castle Rock, Centennial, Elbert, Elizabeth, Ellicott, Englewood, Florissant, Fort Collins, Fountain, Franktown, Larkspur, Littleton, Midfield Village, Monument, Parker, Peyton, Rangely, Sedalia, U.S. Air Force Academy, Widefield, Woodland Park and Colorado Springs. For the first time, foreign students came to participate in the CMO. Three students and their parents traveled all the way from the Philippines to compete: they are Timothy Lim Uy, Gary Anthony Auyong Wu, and Raimund Christopher Buendia Cuesico.

Golden middle schooler Bryce Herdt

In the last year of the XX century, the year 2000, Bryce Herdt received the Olympiad's Young Mathlete Award: then seventhgrader, he achieved the highest result among middle school contestants. This time Bryce, an eighthgrade student in a class of Mrs. Loretta Chandler in Irving Middle School of District 11 of Colorado Springs, received the highest award of CMO, first prize. He was the youngest ever first prize winner in the 18-year history of CMO, and may remain the youngest ever winner. Bryce received the gold medal, the software MATHEMATICA donated by Wolfram Research from Silicon Valley, a graphing calculator donated by CASIO, the set of 4 books by Alexander Soifer, and a $2,000 scholarship to be used at any accredited American university or four-year college.

Bryce became the most amazing winner. This eighth-grade middle schooler suffered from Asperger's Syndrome, an autism spectrum disorder. Sometimes I hear from "normal" folk that being autistic is good for doing math. Such an "opinion" may come only from ignorance and, perhaps, desire to explain own failure. Bryce rejected excuses for failure and succeeded against all odds. His victory was nothing short of heroic. Read his mother Roberta A. Crownover's May 1, 2000 letter to me (sent a year earlier, when we presented Bryce with first honorable mention and Young Mathlete Award):

Dear Dr. Soifer,

I would like to thank you again for all your work and commitment. The Math Olympiad is an incredible event and its worth grows exponentially when applied to Bryce. As a child diagnosed with Asperger's Syndrome, Bryce fights a battle every day that most of us cannot even begin to imagine. Being a part of this wonderful community of math "geeks" gives him a sense of belonging that is precious. For this above all, I thank you...

Thank you again.

Monday, May 7, 2001, 9 to 11 A.M., the Colorado State Senate brought in the CMO-2001 young winner Bryce Herdt with his parents, teacher, principal and superintendent to the floor of the Senate for a special recognition. This was the first ever recognition of a young mathematician in such an important chamber of power.

Second prize was awarded to Mark Pond, a junior from Smoky Hill High School, Aurora. He received the silver medal of the Olympiad, the software MATHEMATICA, a CASIO graphing calculator, a $1,500 scholarship, the book *Geometric Etudes in Combinatorial Mathematics* by V. G. Boltyanski and A. Soifer, and Wolfram Research memorabilia.

Third prizes were presented to two competitors: Ryan Gardner a senior from Ponderosa High School; and Steven Brodhead a junior from Air Academy High School. Each of them received the bronze medal of the Olympiad, a $1,000 scholarship, the software MATHEMATICA, a CASIO graphing calculator, the Boltyanski–Soifer book, and Wolfram Research memorabilia.

The judges have decided to award 6 fourth prizes this year, and also 15 first honorable mentions and 42 second honorable mentions.

A Literary Award was presented – again – to James Carroll, a senior from Arapahoe High School. He was the winner of this award in 1999:

The Brightest Sailor
By James Carroll

Lost at sea
I began to doubt
The storm allowed
No visible way out.

But I remembered
My lessons of yore –
How I could reach
That unseen shore.

The shape of the lake
Mattered not at all,
I knew if I thought,
My flag would not fall –
Down to the level of
Well-past half-mast,
I will conquer this lake at last!

The prize fund of the Olympiad had been generously donated by Ingersoll-Rand Company, Wolfram Research, Inc., CASIO, Air Academy School District 20, Harrison School District 2, Colorado Springs School District 11, Rangely School District RE-4, Douglas County School District RE-1, St. Mary's School, Elbert High School, Fort Collins High School, Rangeview High School of Aurora, Arapahoe High School, Colorado College, UCCS Academic Affairs Vice Chancellor's Office, and UCCS Bookstore.

The Award Presentation Program included *Review of Solutions of the Olympiad Problems* by Alexander Soifer and his lecture *Mathematical Education: proTEST & conTEST* giving a (critical) assessment of President George W. Bush's and Governor Bill Owens's reforms and the role of testing in teaching and learning.

The following guests of honor, hosts, and sponsors addressed the winners and present the awards: Linda Bunnell Shade, Chancellor; Elizabeth S. Grobsmith, Dean, College of Letters, Arts and Sciences; David Nelson, Dean of the School of Education, - all from UCCS; James A. Null, Colorado Springs Councilman; Gregory C. Hoffman, Director, Human Resources, Ingersoll-Rand Company, Colorado Springs; and Paul R. Wellin, Director of Corporate and Academic Affairs, Wolfram Research Inc., Chico, CA, who came from California especially for this ceremony.

In the 18 years of the Colorado Mathematical Olympiad, some 12,500 students have participated during 1984-2001. They have written some 6,000 essays, and have been awarded over $135,000 in prizes.

Days after the Award presentation, on May 11, 2001, my Dean of the College of Letters, Arts and Sciences, Elizabeth Grobsmith decided to confiscate a tiny storage room where the Olympiad securely kept prizes and archives. Of course, the dean had the authority to do that, but she wanted *me* to support the act that would have made running the Olympiad very difficult. The dean must have not read Vladimir Nabokov's masterpiece *Invitation to a Beheading*. There Cincinnatus C., sentenced to death, is expected to help his own executioner in his own execution.

I refused to be Cincinnatus C. and appealed the dean's decision to the Chancellor of my campus, Linda Bunnell Shade. This chancellor distinguished herself by not allowing faculty and staff to even make appointments to see her. Nevertheless, she met with me to articulate the university lawyer's advice: "the board of regents of the University of Colorado pledged $3,000 a year for your running the Olympiad and related travel – that would be provided even if you were to move the Olympiad away from my university, for example, to the [private] University of Denver," said the chancellor, and added" "However, we are not obligated to provide you with a storage room."

Indeed, I was ready to move the Olympiad, for example, to the University of Denver. But strange things happened over the summer of 2001: both Dean Grobsmith and Chancellor Bunnell Shade suddenly resigned! The Interim Dean Tom Christensen and the Interim Chancellor Pam Shockley Zalaback were enthusiastic supporters of the Olympiad. The thunderstorm was over; the Colorado-blue skies once again rose over the Olympiad.

Did I tell you, my city of Colorado Springs has for decades been the capital of the U.S. Olympic movement? Yes, on Boulder Street we have their huge compound, with the Olympic headquarters of the USA, training facilities, and living quarters for the American Olympians. Consequently, we at my university get the Olympians in our classes. (I too have had among my students my share of the champions.) My university established an *admissions window* for the Olympians, i.e., exceptions in admission requirements especially for the Olympic sportsmen. For years, I asked for the same admissions window for the Colorado Mathematical Olympiad winners. I argued: "we ought to reward heads as much as we reward arms and legs." Of course, this was a joke: it takes not only arms and legs but also guts and brains to win in Olympic Games. In looking through the

Olympiad's archive, I came across one of my pleas, dated March 27, 2001 and addressed to the Vice Chancellor for Academic Affairs John Pierce:

Dear Vice Chancellor Pierce:

University of Colorado at Colorado Springs' (UCCS) admission rules do not seem to serve their purpose of selecting the best and brightest students. Indeed, one can make an argument that talent may be a liability at UCCS admissions. Matt Kahle's example tells it all. With 2.0 GPA, he was rejected by UCCS from becoming our freshman. And now Matt has been invited, with most attractive scholarships, to such premier Ph.D. programs in Mathematics as University of California, Berkeley; University of Washington, Seattle; and University of Waterloo, Ontario. And it is not like UCCS did not know whom they were rejecting: I wrote to Admissions that Matt won First Prize in 1990 and 1991 Colorado Mathematical Olympiad, and UCCS needed this most gifted Colorado mathematician of 1989-1991 much more than Matt needed UCCS.

It is time for UCCS to stop basing its admission decisions on banal numbers alone, such as GPA, SAT, ACT, KGB, etc. I would favor giving minimal – if at all – weight to these stupid standardized tests, and put our emphasis on talent. Admission interviews and essays ought to be implemented and given major weight. Talent must be treasured instead of being ignored.

Other schools understood it. Stanford admitted Travis Kopp, the 1993 and 1994 Colorado Mathematical Olympiad winner right from 11th grade, without even a high school diploma. Duke University offers admission and scholarships to all winners of USA Mathematics Olympiad. University of California President recently promised to downplay standardized tests in their admission process.

The whole UCCS Admissions process must be overhauled, taken off pedestal and put to good use of bringing the best talent into our university.

As a first small step, I hereby request UCCS to admit everyone who won First, Second or Third prize in Colorado Mathematical Olympiad (CMO) – regardless of their GPA, SAT, ACT. Such an achievement in a statewide competition of 400-1000 State's best students (attained by 3 to 10 students a year) proves talent and promise of success beyond any doubt. Moreover, this policy must not only be adapted, but also advertised. CMO winners, with their Prizes, ought to receive letters of invitation to apply to UCCS with guaranteed admission. Surely, we would not get all CMO winners at UCCS – Harvard must get its share – but we would get some. We would also make an important statement by such a policy, that talent matters in the world in general and at UCCS in particular. And we would eliminate shameful stories like Matt Kahle's rejection.

You may ask me how CMO's past winners fared. The 1984 winner Russel Shaffer went to MIT and then got Ph.D. in theoretical computer science from Princeton on a full NSF scholarship. The 1985 winner Richard Wolniewicz got Ph.D. in computer science from [CU] Boulder, and is a co-owner of his own software company. The 1986-1988 triple winner David Hunter went to Princeton then earned a Ph.D. in Statistics from University of Michigan, Ann Arbor and is now an Assistant Professor. The 1988 co-winner Gideon Yaffe first graduated from Harvard in Mathematics and Drama, and last I heard was working on a Ph.D. in Philosophy at Berkeley. The 1990-1991 winner Matt Kahle is going to the University of Washington's Ph.D. program in Mathematics on full scholarship.

As you can see, CMO has a record to be proud of. Moreover, CMO has been held at UCCS for about 18 years. It is only natural for CMO's host, UCCS, to try and recruit some of the winners.

I got no reply from Vice Chancellor Pierce. However, with new wonderful Chancellor Shockley Zalaback and Dean Christensen my long standing request was accepted and on September 26, 2001 signed onto the three-way agreement between the dean, director of admissions and records and me. From now on, it sufficed to win one medal – gold, silver, or bronze – over all middle and high school years, to get exactly the same admissions window that has existed for the U.S. Olympic athletes!

Moreover, Chancellor Pam Shockley Zalaback asked me "what do you think about my granting $1000 to each Colorado Mathematical Olympiad's medalist, who enrolls in our university as a new freshman?" "I am surprised no one offered this earlier," was my reply. The chancellor and I signed the Scholarship plan on February 7, 2002. This is how "The Chancellor's Scholarships for the Olympiad's Medalists" were born!

In July 2001 I had the pleasure of serving as a Coordinator of the International Mathematical Olympiad (IMO) in Washington D.C., and visit with my American friends Ron Graham, Murray Klamkin, George Berszenyi and others; and with my World Federation friends Peter Taylor, and Bill and Krys Richardson (sadly, it was the last time I saw her). China won, as usual, with the USA and Russia tying for second place.

Problems 18

Problems 18.1, 18.2, and 18.3 were created especially for this Olympiad by Alexander Soifer. Problem 18.4 is A. Soifer's generalization of an old 8×8 chessboard question. Problem 18.5 came from the Russian mathematical folklore. These problems were selected and edited by the Problem Committee: Gary Miller and Alexander Soifer.

18.1 2001: A SPACE ODYSSEY (*A. Soifer*). HAL-9000 and Dave Bowman decide to resolve in a civilized way who controls the spaceship *Discovery*. They place 2001 computer chips equally spaced along the edge of a round table, and in turn take off the table any number of consecutive chips between 1 and 21. The one who takes the last chip wins control of the spaceship. Find a strategy that allows HAL or Dave to win control regardless of how the opponent may play. Oh, yes: HAL goes first.

18.2 POSITIVE^2 (*A. Soifer*). Is there a way to fill a 2001×2001 square table with pluses and minuses such that no series of interchanging all signs in any 1000×1000 or 1001×1001 square of the table would produce a table with all pluses?

18.3 BORDERLINE POSITIVE (*A. Soifer*). All numbers of a 2001×2001 square table of real numbers add up to a positive sum. Is there a way to transform the table into a table with positive sum of all numbers along its perimeter by a series of interchanges of any two columns or any two rows?

18.4 TWO-ON-ONE ATTACK (*A. Soifer*).

(a) Find the smallest number of rooks that must be placed on a 6×6 chessboard so that every square is attacked by at least two rooks. (We say that a square is "*attacked*" by a rook if the square and the rook are in the same row or column of the chessboard.)
(b) Solve the same problem for a 2001×2001 chessboard.

18.5 SURVIVAL OF THE BRIGHTEST. A sailor was caught in a storm with no visibility and enough fuel for only 162 miles of travel in a 2001 square-mile lake that has no islands. He can move along straight lines only, but change direction any time he pleases. Prove that there is a strategy that allows the sailor to reach the shore and survive.

Solutions 18

18.1. Dave Bowman wins. HAL makes his first move taking x consecutive chips off the table. Then on his first move, Dave takes chips off the side of the table diametrically opposite to HAL's move, and, moreover, Dave takes a quantity of the parity opposite to the parity of x. As a result, the circle of chips gets split into two equal length arcs of chips, symmetric with respect to a line, call it L, and thus Dave can now win by playing symmetry, i.e., taking off the table the chips that are symmetric with respect to L to the chips taken off by HAL. ∎

18.2. Having created this problem and a solution (see the second solution below), I felt that another solution was possible using an invariant, but failed to find it. Two days after the Olympiad, the past double-winner of the Olympiad Matt Kahle, who had traveled to Colorado Springs for the weekend to serve as a judge, found the solution that eluded me. It is wonderful and concise, and I will start with it.

First Solution, by Matt Kahle, obtained on April 22, 2001.

Denote the given 2001×2001 table by T, and define $S = \{$the set of all unit squares of T, except those in the middle row$\}$ (Figure 18.1). Observe that no matter where a 1000×1000 square is placed on T, it intersects S in an even number of unit squares, because there are 1000 equal columns in the intersection. Observe also that no matter where a 1001×1001 square is placed on T, it also intersects S in an even number of unit squares, because there are 1000 equal rows (one row is always missing, since the middle row is omitted in S).

We can now create the required assignment of signs in T that cannot be converted into all pluses. Let S have any assignment with an *odd* number of $+$ signs, and let the middle row missing in S be assigned signs in any way. No series of operations can change the parity of the number of pluses in S. We are done. ■

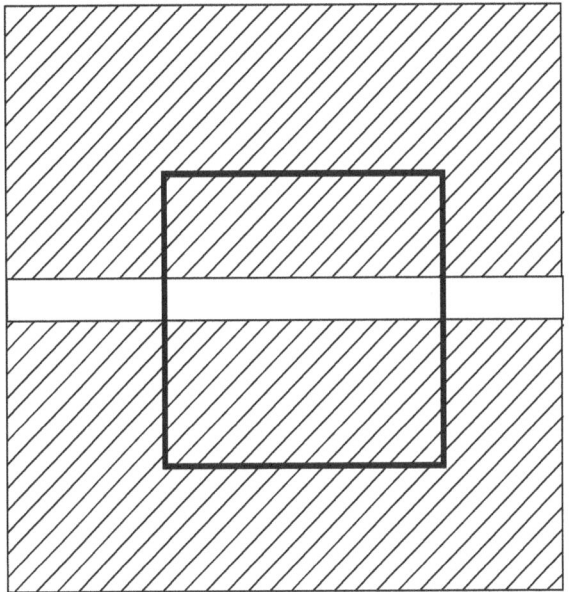

Fig. 18.1

Second Solution. Here is my solution, which does not require the insight of the previous one. Let us pick one particular 1001×1001 square T_{1001} of the given table T. The square T_{1001} works like a *light switch*: the first time we interchange signs, but when we apply the operation twice, the signs return to the original ones. Thus, each 1000×1000 and 1001×1001 square of the table serves as a light switch. How many light switches do we have?

In order to define one 1001×1001 square, it suffices to define the location of its top row and leftmost column. Any one of the first 1000 rows of the table counting from above can serve as a top row (we need room for 1000 more rows below the top row to fit the 1001×1001 square). Similarly, any one of the first 1000 columns of the table counting from the left can serve as the leftmost column of the 1001×1001 square. Thus, T contains $1000 \times 1000 = 1000^2$ distinct squares of the size 1001×1001.

Similarly we can observe that T contains $1001 \times 1001 = 1001^2$ squares of size 1000×1000. Thus, the table T has $1000^2 + 1001^2$ switches. If we start with a table T^+ of all pluses, by a series of allowed operations, we can produce $2^{1000^2 + 1001^2}$ distinct tables. But the total number of ways to fill a 2001×2001 table with pluses and minuses is much greater: 2^{2001^2}. Thus, there is a way T' to fill the table, such that it cannot be obtained from the table T^+ of all pluses by a series of allowed operations. Then, of course, T^+ cannot be obtained from T' either. Thus, T' is the required filling of the table with pluses and minuses. ∎

Problem 18.2 was not solved by any of the Olympians. I see it now as my fault: this beautiful problem was too hard to be the second problem in our set of five problems of increasing difficulty.

18.3. A series of interchanges of two columns or two rows can produce many perimeter sets of numbers. Let us add together perimeter sums of all possible perimeters, call this sum S. Observe that by an appropriate interchange of columns and rows, we can interchange any entry of the table with any other entry. Therefore, S is the sum of the identical number C (C is a positive integer of course) of each entry of the original table.

We have

$$P_1 + P_2 + \ldots + P_n = S = C(a_{1,1} + a_{1,2} + \ldots + a_{2001,2001}).$$

But the sum of all entries of the given table is positive: $a_{1,1} + a_{1,2} + \ldots + a_{2001,2001} > 0$. Therefore,

$$P_1 + P_2 + \ldots + P_n > 0,$$

and thus at least one of the summands P_i in the left side above is positive. But P_i is the sum of all numbers along one perimeter. We are done.

Remark. It really is irrelevant to know the exact value of C. But to satisfy your curiosity, I can calculate it – especially since the result is nice. Let us do it for a $n \times n$ table (in our problem $n = 2001$).

I claim that

$$C = \binom{n-1}{1}\binom{n}{2} + \binom{n-1}{1}\binom{n}{2} - \binom{n-1}{1}\binom{n-1}{1} = (n-1)^3 \tag{*}$$

Indeed, a perimeter sum is the sum of all elements of two rows and two columns minus 4 elements at their intersection, because we are counting these 4 elements twice in adding two rows and two columns.

How many times does a row appear in the sum S of all perimeters? Fix a row; there are $\binom{n-1}{1}$ ways to choose another row of the perimeter. The pair of rows can be combined with each pair of columns, and there are, of course, $\binom{n}{2}$ such pairs. Thus, we see that each row will appear $\binom{n-1}{1}\binom{n}{2}$ times. A similar argument computes the number of times each column is repeated in C. This explains the first two summands in equality (*). Now we need to subtract twice-counted perimeters' corner elements. Fix an element. How many ways are there to have it as a corner of a perimeter? We have to choose the opposite corner, and for that we have to choose its row and column. There are $\binom{n-1}{1}$ choices for each of them, and thus $\binom{n-1}{1}\binom{n-1}{1}$ computes the amount we need to subtract in (*). ∎

18.4(a). It is easy to see that 12 rooks suffice for a double attack of every square of a 6×6 chessboard: just place the rooks on two main diagonals of the board. We can actually do much better. Figure 18.2 shows that in fact 8 rooks suffice.

Let us now show that 8 is the smallest possible number of rooks for the task of double-attack. Assume that n rooks are placed on the board and attack every square twice, and $n \leq 7$.

The property of a rook to attack or not to attack a square does not change if we were to interchange any two columns or any two rows of the board. Let us build the longest possible main diagonal of rooks by interchanging rows and columns (Figure 18.3).

Observe that there are no rooks in the $r \times r$ lower right square S, for otherwise we would be able to lengthen the longest possible diagonal of rooks. Now look at the (empty) squares of the main diagonal inside S; they are marked by small dots in (Figure 18.3). Each of these

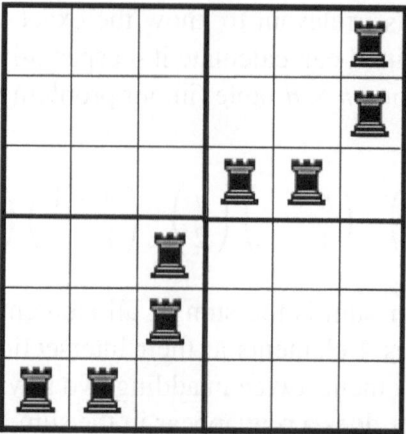

Fig. 18.2

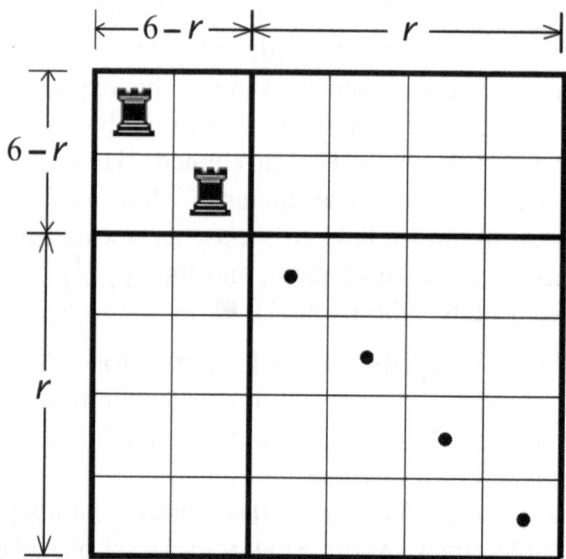

Fig. 18.3

squares must be attacked twice, but one rook positioned outside of S can attack at most one "dotted" square. Therefore, there are at least $2r$ rooks altogether in the two $(6 - r) \times r$ rectangles of the board. Thus, we have accounted for at least $(6 - r) + 2r = 6 + r$ rooks.

Now observe that $r \leq 1$, for otherwise $r \geq 2$ and we would account for at least $6 + r \geq 8$ rooks, in contradiction to our assumption that we have $n \leq 7$ rooks.

Thus, our diagonal of rooks is fairly long: we have $6 - r \geq 5$ rooks on it. Each of these diagonal rooks' squares is attacked by the rook standing on it, but must be attacked once more. But one rook can attack at most 2 rooks of the main diagonal, therefore, we can account for at least $n \geq 5 + \frac{5}{2} = 7.5$ rooks, which contradicts our assumption that $n \leq 7$, and we are done. ■

18.4(b). How does one generalize such a result as 8 rooks for the 6×6 board? What is 8 with respect to 6? Well, it is $6 + 2$. It is also $6 \times \frac{4}{3}$. Figure 18.2 suggests a linear relationship. Indeed, by generalizing the configuration of Figure 18.2 we can achieve the task on an $n \times n$ chessboard by placing $\frac{4}{3}n$ rooks (in our problem 18.4 both numbers, 6 and 2001, are divisible by 3).

Let us now show that $\frac{4}{3}n$ is the smallest possible number of rooks for the task of double-attack. Assume that m rooks are placed on the board and attack every square twice, and $m < \frac{4}{3}n$.

The property of a rook to attack or not to attack a square does not change if we interchange any two columns or any two rows of the board. Let us build the longest possible main diagonal of rooks by interchanging rows and columns (look at the same Figure 18.3; just mentally replace "6" by "n").

Observe that there are no rooks in the $r \times r$ lower right square S, for otherwise we would be able to lengthen the longest possible diagonal of rooks. Now look at the (empty) squares of the main diagonal inside S marked by dots in Figure 18.3. Each of these "dotted" squares must be attacked twice, but one rook positioned outside of S can attack at most one dotted square. Therefore, there are at least $2r$ rooks altogether in the two $(6 - r) \times r$ rectangles of the board. Thus, we have accounted for at least $(6 - r) + 2r = n + r$ rooks.

Now observe that $r < \frac{n}{3}$, for otherwise we would account for at least $n + r \geq \frac{4}{3}n$ rooks, in contradiction to our assumption that $m < \frac{4}{3}n$. Thus, our diagonal of rooks is fairly long: $n - r \geq \frac{2}{3}n$. Each of these diagonal rooks' squares is attacked by the rook standing on it, but must be attacked once more. Each of the $2r$ rooks we accounted for in the two $(n - r) \times r$ rectangles of the board, can attack at most one diagonal rook, leaving $(n - r) - 2r = n - 3r > 0$ diagonal rooks to "seek" the second attack. But any one rook can attack at most two rooks of the main diagonal, therefore, we can account for at least $\frac{n-3r}{2}$ more rooks, bringing the total number of rooks to at least

$$n + r + \frac{n - 3r}{2} = \frac{3n - r}{2} > \frac{\frac{8}{3}n}{2} = \frac{4}{3}n.$$

This contradicts our assumption that $m < \frac{4}{3}n$, and completes the proof that we need at least $\frac{4}{3}n$. The construction in Figure 18.4 demonstrates that $\frac{4}{3}n$ rooks suffice.

Now we have to substitute $n = 2001$ to get $\frac{4}{3}(2001) = 2668$.

Remark. You may wonder how many rooks are needed for a double-attack on an $n \times n$ chessboard if n is not divisible by 3. The answer is $\lceil \frac{4}{3}n \rceil$, where $\lceil a \rceil$ denotes the smallest integer greater than or equal to a. The proof that a lesser number of rooks cannot do the job is the same as above. I invite you to try to construct compositions of $\lceil \frac{4}{3}n \rceil$ rooks that do the job, and only then look at my constructions in Figures 18.5 and 18.6. ∎

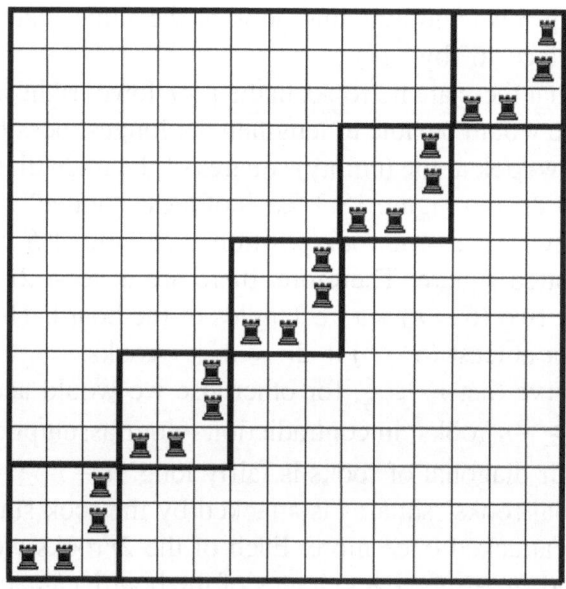

Fig. 18.4

None of the Olympians solved Problems 18.4(A) or 18.4(B).

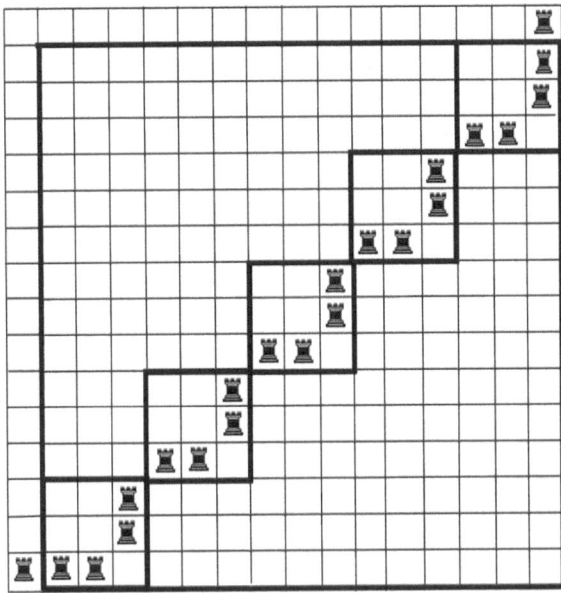

Fig. 18.5

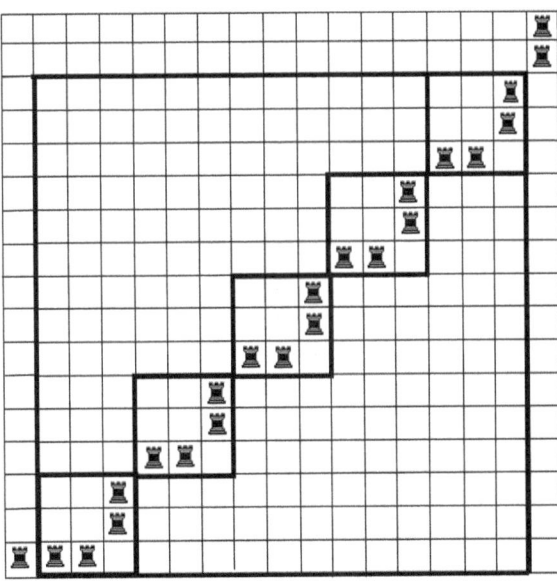

Fig. 18.6

18.5. The circumference of a circle of area 2001 is equal

$$2\pi \sqrt{\frac{2001}{\pi}} = 2\sqrt{2001\pi} \approx 158.6.$$

We can therefore circumscribe a regular n-gon P of perimeter at most 162 about this circle for an appropriately chosen n. Such a P would obviously have area strictly greater than 2001.

The strategy is now clear. The sailor puts the polygon P (made to scale :-) on his map in such a way, that P goes through the point where the sailor is located. Then the sailor travels along the perimeter of P. He will hit the shore before coming back to the starting point, for otherwise he would travel all the way around P, which thus would be contained completely inside the lake. But then the lake would have area greater than 2001 since P is of area greater than 2001, thus contradicting the conditions of the problem. ∎

Nineteenth Colorado Mathematical Olympiad
April 19, 2002

Historical Notes 19

Bryce Herdt wins again

The nineteenth Colorado Mathematical Olympiad (CMO-2002) took place on April 19, 2002, the day before the third anniversary of the Columbine tragedy. It brought together 614 middle and high school students (a 28% increase from the previous year). Contestants came from all over Colorado: Agate, Aurora, Benett, Black Forest, Brighton, Canon City, Cascade, Centennial, Colorado Springs, Deer Trail, Denver, Dinosaur, Divide, Englewood, Falcon, Fort Collins, Fort Lupton, Fort Morgan, Fountain, Fowler, Las Animas, Littleton, Luatkins, Manitou Springs, Monument, Pueblo, Rangely, Weldona, Widefield, and Woodland Park. The organizers were unable to explain how kids from Bagg, Wyoming, slipped into the competition :-).

The Awards Presentation opened with the words of the Governor of the State of Colorado Bill Owens, which I read for him:

> On behalf of the citizens of the State of Colorado, it is my pleasure to welcome you to the award presentation ceremony of the 19th Annual Colorado Mathematical Olympiad, and to congratulate all of the participants who took part in the Olympiad for a job well done. We in Colorado treasure your achievements every bit as much as the accomplishments of others involved in athletics and other extra-curricular activities. Your example will undoubtedly inspire all Colorado students to strive for greater success.
>
> Education is my top priority in Colorado. In our day and age of high technology and global communication everybody needs to

A. Soifer, *The Colorado Mathematical Olympiad and Further Explorations: From the Mountains of Colorado to the Peaks of Mathematics,* DOI 10.1007/978-0-387-75472-7_33, © Alexander Soifer, 2011

be proficient in mathematics. Your hard work and talent promises a bright future to you, and thus to all of us. We live in an exciting time with tremendous opportunities for new discoveries and knowledge. With your curiosity and dedication to learning in general, and mathematics in particular, I am confident in your success.

I thank you and wish you every success.

Bryce Herdt is golden again!

The judges awarded two first prizes: for the second year in a row to Bryce Herdt, now a freshman from Mitchell High School, and to Steven Biles, a junior from Palmer High School. Each of them received the gold medal of the Olympiad, the software MATH-EMATICA donated by Wolfram Research from Silicon Valley, a graphing calculator donated by CASIO, and a $2,000 scholarship to be used at any accredited American university or four-year college. Steven also received a set of four books by Alexander Soifer, and Bryce four other books (since he had won Soifer's books in 2001). Do not forget Bryce's name – he will be back as a sophomore!

Second prize was awarded to Tim Morley, a junior from James Irwin Charter School. He received the silver medal of the Olympiad, the software MATHEMATICA, a $1,000 scholarship, and a book *Geometric Etudes in Combinatorial Mathematics* by V. Boltyanski and A. Soifer.

Third prizes were presented to the following six Olympians: Autumn Petros-Good a freshman, Curtis Larimer a sophomore, Jeff Graham, a junior, and Joseph Koehler, a junior – all from Palmer High School; Devon Horntvedt, a senior from Rangely High School; and Mark Pond, a senior from Smoky Hill High School. Each of them received the bronze medal of the Olympiad, the software MATHEMATICA, CASIO or TI graphing calculator, and the Boltyanski–Soifer book.

UCCS Chancellor Pamela Shockley-Zalaback and Superintendent of Colorado Springs School District 11 Norman Ridder

UCCS Chancellor Pamela Shockley-Zalaback instituted "The Chancellor's Scholarship for CMO Medalists" effective with this year's Olympiad. As we learned in the previous *Historical Notes*, from this time on, each gold, silver and bronze medalist would be awarded an additional $1000 upon application and acceptance to the University of Colorado at Colorado Springs as a new freshman. Pam's act was a major contribution to education of the young gifted mathematicians of Colorado, and I was most grateful to Chancellor Shockley-Zalaback for her wisdom and generosity. It was entirely her initiative – I did not ask for these scholarships!

The judges decided to award 4 fourth prizes this year, 53 first honorable mentions, and 55 second honorable mentions.

All 160 winners received Wolfram Research's mathematical T-shirts.

A Literary award was presented to Mike Gonzales, a senior from Smoky Hill High School near Denver, for his humorous poem on *King Arthur and the Knights of the Round Table* problem:

At the long table the knights would eat their food
But could not resist to fight and feud
Young Arthur said, "to maintain order
Let's get a table without one corner."

"A round table" said one "That's got a nice ring,"
"That brilliant lad, let's make him king!"
So off they went for a table a thousand foot round
That cost taxpayers many a pound.

Then brought it back to the castle door
It would not fit, were they to sit on the floor?
Arthur realized that he was young
Perhaps next time he would bite his tongue.

But it was too late – he had not foreseen
One mistake as king would bring him to the guillotine!

The prize fund of The Olympiad was generously donated by Ingersoll-Rand Company, Wolfram Research, Inc., CASIO, Texas Instruments, Air Academy School District 20, Harrison School District 2, Colorado Springs School District 11, Bennet School District 29 J, St. Mary's School, Sand Creek High School, Mann Middle School, Arapahoe High School, Fort Collins High School, Rangely School District RE-4, Colorado College, CU-Colorado Springs Vice-Chancellor's Office, and CU-Colorado Springs Bookstore. Widefield School District 3 did not contribute; in general this district has been an on-again off-again sponsor.

The Award Presentation Program included *Review of Solutions of the Olympiad Problems* by Alexander Soifer and his lecture *The Games We Play.*

The following guests of honor, hosts, and sponsors addressed the winners and present the awards: Pamela Shockley, Chancellor; Thomas Christensen, Dean, College of Letters, Arts and Sciences – both from CU-Colorado Springs; James A. Null, Colorado Springs City Councilman; Gregory C. Hoffman, Director, Human Resources,

Ingersoll-Rand Company, Colorado Springs; and James Carroll, a freshman at Brigham Young University and a former Olympiad winner of two literary awards and one third prize. You will meet James again in Part V.

Problems 19

All problems were created by Alexander Soifer especially for this Olympiad. Problem 19.5 was inspired by the celebrated Norwegian mathematician Øystein Ore's 1960 paper. The problems were selected and edited by the Problem Committee: Dr. Col. Bob Ewell, Gary Miller, and Alexander Soifer.

19.1 WHAT'S HIDING IN THE CORNER? (*A. Soifer*). Each square of a 2002×2002 chessboard contains a non-zero number such that no matter where on the board we put 2002 rooks that do not attack each other, the product of the 2002 numbers covered by the rooks stays the same. The upper left corner contains 1, the upper right corner contains 14, and the lower left corner contains 143. What number is contained in the lower right corner?

   (We say that two rooks attack each other if they are in the same row or column of the board.)

19.2 A FITTING QUESTION (*A. Soifer*). A cross on a chessboard consists of a unit square together with its four neighbors above, below, to the right, and to the left. Find the largest number of crosses that can fit without intersection on a regular 8×8 chessboard.

19.3 PROBLEM BLUE (*A. Soifer*). Each unit square of a 2002×2002 chessboard is colored in one color, red or blue, in such a way that no matter where we place a 3×3 square along the lines of the board it would contain at most five red squares. What is the smallest number of blue squares the entire board can have?

19.4 THE 1-10-100 PROBLEM (*A. Soifer*). Given 100 10-element sets such that every two sets have precisely 1 element in common, prove that there is an element that is shared by all sets.

19.5 KING ARTHUR AND THE KNIGHTS OF THE ROUND TABLE (A. Soifer).

(a) Each of the 2002 knights (including King Arthur!) got into a feud with up to 1000 of the other knights. Prove that King Arthur can always seat all 2002 knights at the Round Table so that no feuding pair of knights sits together.

(b) Show that even King Arthur may fail to seat the knights peaceably (as required in part a) if each knight were to feud with 1001 of the other knights.

Solutions 19

19.1. Place 2002 rooks on the main diagonal of the chessboard, which goes from the square containing the number 1 to the square with the number we want to determine, call it x (Figure 19.1). The 2002 rooks do not attack each other.

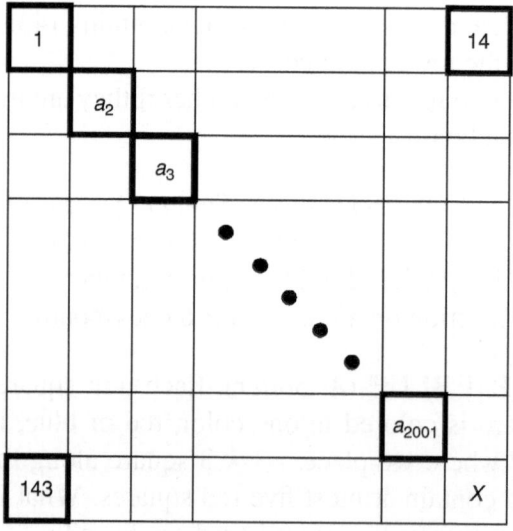

Fig. 19.1

We can now move the two rooks covering 1 and x to cover 14 and 143 while keeping the remaining 2000 rooks in their places on the main diagonal. The 2002 rooks in the new position do not attack each other. Therefore, the product of the numbers under the rooks in the first position and the second position are equal:

$$1 \times a_2 \times a_3 \times \ldots \times a_{2001} \times x = 14 \times a_2 \times a_3 \times \ldots \times a_{2001} \times 143$$

Therefore, $1 \times x = 14 \times 143$, and $x = 2002$. ■

19.2. Observe that there are as many crosses on the board as their centers. The center cannot be in any perimeter square. In order to cover a square in a perimeter row R_1, the cross's center must lie in the row R_2 adjacent to R_1 (Figure 19.2). Every cross with its center in R_2 covers three squares of R_2. Therefore, there are at most two cross centers in R_2. Thus, at most two squares of row R_1 can be covered by crosses! Thus, at least six squares of R_1 are uncovered. Accounting for four perimeter sides of the given 8×8 chessboard, we detect at least $24 - 4 = 20$ empty perimeter squares (I subtracted 4 because the corner squares of the perimeter were counted twice). But then the crosses can cover at most $8 \times 8 - 20 = 44$ squares, and thus we can fit at most $\lfloor 44/5 \rfloor = 8$ crosses (here the symbol $\lfloor a \rfloor$ denotes the largest integer that is not greater than a).

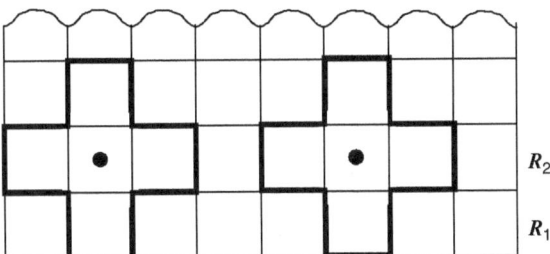

Fig. 19.2

On the other hand, the whole plane can be tiled by crosses (Figure 19.3), and an appropriate choice of the 8×8 square on the tiled plane illustrates that eight crosses can be fitted (Figure 19.4). ■

19.3. When I think of five unit squares inside a 3×3 square S, I see the union of one red row and one red column of the 3×3 square – any row and any column. This thought suggests the following coloring: we color red the 1st, 4th, 7th, ..., 2002nd columns, and we color red the

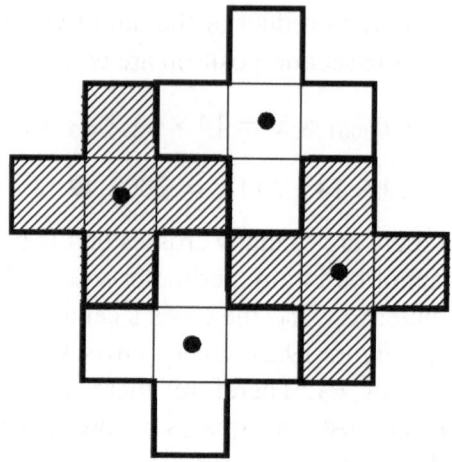

Fig. 19.3

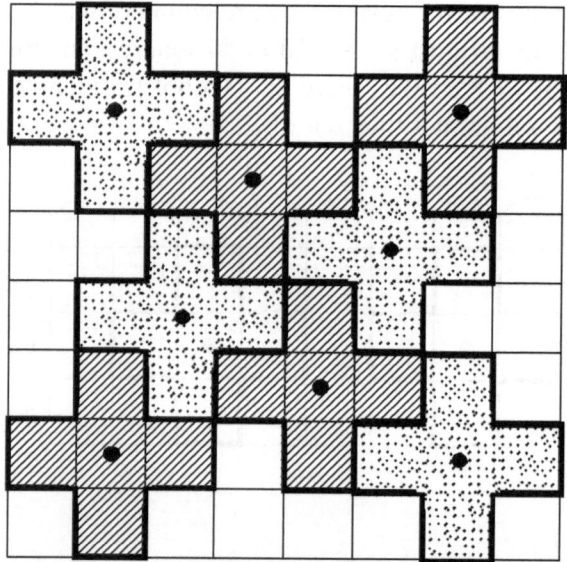

Fig. 19.4

$1^{st}, 4^{th}, 7^{th}, \ldots, 2002^{nd}$ rows of the chessboard, leaving the rest blue (Figure 19.5).

This is a desired optimal coloring! Indeed, no matter where we place a 3×3 square along the lines of the board it contains exactly five red squares. Furthermore, Figure 19.5 presents a partition of the

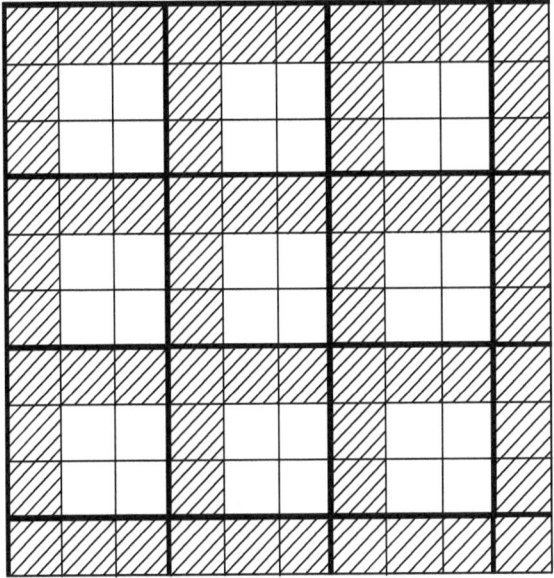

Fig. 19.5

2002×2002 chessboard into disjoint 3×3 squares plus the last row and the last column. We have colored the maximum allowed five squares red in each of the 3×3 squares, and we colored the entire last row and the last column red – in view of the fact that our areas are disjoint and we got the maximum in each of them, we have produced the maximum number of red squares, and thus minimum number of blue squares. And this blue minimum is equal to

$$4(2001/3)^2 = 1,779,556. \ \blacksquare$$

19.4. Pick one of the given 100 sets $A_1 = \{a_1, a_2, \ldots, a_{10}\}$. There is one of the elements $a_1, a_2, \ldots, a_{10}$ that is shared by A_1 with at least 10 more sets, for otherwise a_1 would be shared by A_1 and at most 9 other sets; a_2 would be shared by A_1 and at most 9 other sets; etc.; a_{10} would be shared by A_1 and at most 9 other sets – and the total number of given sets would be at most $1 + 10 \times 9 = 91$ sets, whereas we were given 100 sets. Thus, one element of A_1, say a_1, is shared by A_1 with 10 more sets $A_2, A_3, \ldots, A_{11}$.

Pick any of the remaining sets, call it B. Then B must have an element b_i in common with the set A_i for each $i = 1, 2, \ldots, 11$.

Moreover, all eleven elements b_i must be distinct, for sets $A_1, A_2,$ $A_3, \ldots, A_{11}$ can have only one element in common, and this element is a_1. Thus, we have found an 11-element set B in contradiction to the fact that all given sets consisted of 10 elements.

Figure 19.6 well illustrates our analytical findings.

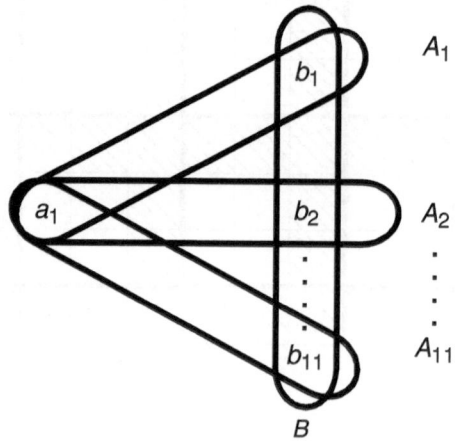

Fig. 19.6

This problem was inspired by the Helly Theorem. In turn I would like to continue this theme with you in *Further Exploration* E18 dedicated to *Three Classic Theorems of Mathematics.* ∎

19.5 (a) It is convenient to represent each knight by a vertex (of arbitrary location on the plane): if a pair of knights is in a feud, we indicate it by connecting the corresponding pair of vertices by an edge (of arbitrary shape). We get what we call the *Graph G of Feuds*. This graph uniquely defines its complement, or what we will call the *Graph $\overline{G}$ of Peace*: we use the same set of vertices in $\overline{G}$ as we did in G and connect by an edge a pair of vertices in $\overline{G}$ if and only if the pair is *not* connected in G. Thus, of course the graph $\overline{G}$ shows us all non-feuding pairs of knights (see an illustration of this duality for a small group of five knights in Figure 19.7).

It is also convenient to talk about the *degree* of a vertex in a graph as the number of lines that emanate from that vertex. We are told that every knight is feuding with up to 1000 other knights. This means

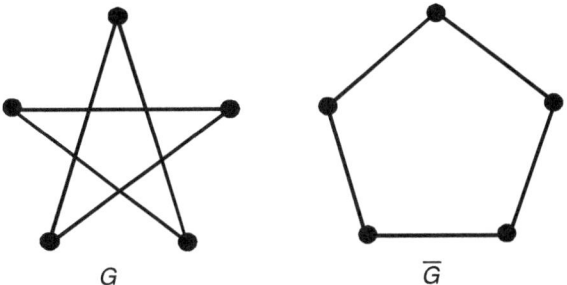

G $\overline{G}$

Fig. 19.7

precisely that the degree of every vertex v in the Graph of Feuds G is at most 1000; we record it as deg $v \leq 1000$. Observe now that if we look at the same vertex in both graphs G and $\overline{G}$, their degrees must add up to precisely 2001 (in both of these graphs taken together, each vertex is connected to all 2001 other vertices!). This implies that in the Graph of Peace $\overline{G}$ every point has degree of at least $2001 - 1000 = 1001$.

In this new language, we have to prove that King Arthur can seat all 2002 knights at the Round Table peaceably – but this means precisely that the Graph of Peace $\overline{G}$ has what is called in Graph Theory a *Hamiltonian cycle*, or a walk along the edges through all the 2002 points that does not repeat a vertex until it ends at the vertex it started from.

OK, if such a Hamiltonian cycle exists in $\overline{G}$, we are done. Otherwise, let us add an edge connecting a pair of vertices in $\overline{G}$. If now we have a Hamiltonian cycle, we stop. If not, we continue adding one edge at a time until a Hamiltonian cycle emerges for the first time. Let $a_1 a_2 \dots a_{2002} a_1$ be this Hamiltonian cycle and $a_{2002} a_1$ be that last added edge that made a Hamiltonian cycle appear (Figure 19.8).

If there is an edge from a_1 to a_i for some i, then there must not be an edge from a_{2002} to a_{i-1}, for otherwise we would get a Hamiltonian cycle $a_1 a_2 \dots a_{i-1} a_{2002} a_{2001} \dots a_i a_1$ (see Figure 19.6), which does not include the edge $a_{2002} a_1$ in contradiction to this edge being the first to make a Hamiltonian cycle possible. Thus, for every edge emanating from a_1, we have an absence of an edge from a_{2002} to another vertex. But this means that the sum of degrees of a_1 and a_{2002} is at most 2001, and this contradicts the fact that the degree of each point in the Graph of Peace $\overline{G}$ is at least 1001. We have proven that the Graph of Peace $\overline{G}$ contains a Hamiltonian cycle! ∎

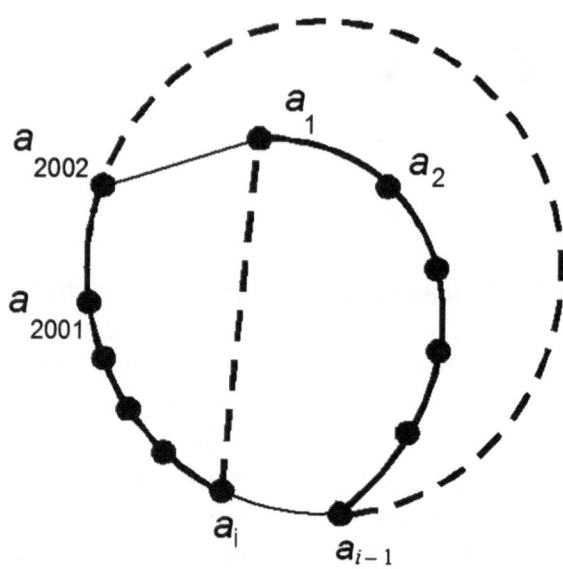

Fig. 19.8

19.5 (b) If each knight feuds with 1001 of the other knights, the degree of every vertex in the Graph of Feuds G is exactly 1001. This implies that the degree of every vertex in the Graph of Peace $\overline{G}$ is precisely $2001 - 1001 = 1000$. We have to construct an example of a Graph of Peace $\overline{G}$ with the degree of each vertex 1000 that does not have a Hamiltonian cycle. It is easy to imagine such a graph $\overline{G}$: we create it out of two components, each consisting of 1001 vertices connected to each other (Figure 19.9).

It is noteworthy to observe that the corresponding Graph of Feuds G displays two armies, say those of Mordred and of King Arthur: every knight of Mordred is in feud with every knight of King Arthur and vice versa (Figure 19.10). Our job is done, and we reluctantly leave Camelot with its King Arthur, Queen Guinevere, Lancelot, Mordred, and other Knights of the Round Table...

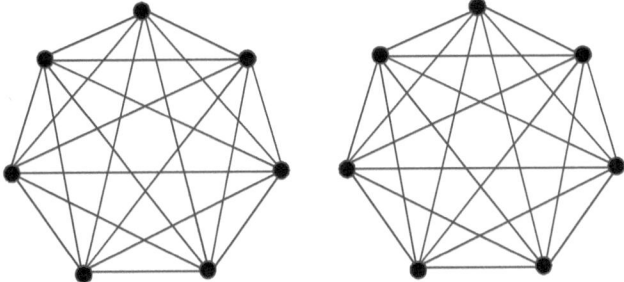

Fig. 19.9

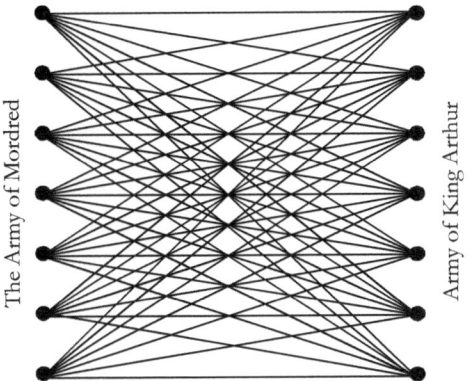

Fig. 19.10

But we will meet again to continue chivalry in *Further Exploration* E19 – see you there!

No one solved Problem 19.5(A), but five Olympians solved Problem 19.5(B). ∎

Twentieth Colorado Mathematical Olympiad
April 18, 2003

Historical Notes 20

Bryce Herdt, now a sophomore, wins for the third straight time!

In December 2002 I left my home in Colorado Springs to serve as a Long Term Visiting Scholar at The Center for Discrete Mathematics and Theoretical Computer Science (DIMACS), a joint project of Princeton University, Rutgers University, and a few top research industrial labs), and as a Visiting Fellow at Princeton University's Mathematics Department. It was a privilege to work in the historic Fine Hall of Princeton-Math, and to befriend senior legendary colleagues, such as Harold W. Kuhn (who nominated the "Beautiful Mind" John Nash for the 1994 Nobel Prize), Edward Nelson (creator of the chromatic number of the plane problem), John H. Conway, Yakov and Elena Sinai, Robert Gunning, the great historian of science Charles Coulston Gillispie, pre-Columbian and ancient art expert and collector Gillette Griffin (in his house, I got to hold in my hands a torso by Michelangelo and drink wine out of fifth century BC ancient Greek cups). I gave talks at the Institute for Advanced Study, Princeton-Math, Rutgers-Math, Claude Shannon Laboratory of AT&T.

As you may understand, I was on sabbatical leave from the University of Colorado. But can the CMO take a sabbatical, and especially a 20-month sabbatical? Of course not! And so I arranged to return to Colorado Springs every April to run both the Olympiad and its Award Presentations.

The Twentieth Colorado Mathematical Olympiad (CMO-2003) took place on April 18, 2003, two days before the fourth anniversary

A. Soifer, *The Colorado Mathematical Olympiad and Further Explorations:*
From the Mountains of Colorado to the Peaks of Mathematics,
DOI 10.1007/978-0-387-75472-7_34, © Alexander Soifer, 2011

of Columbine and in the midst of the war in Iraq. It brought together 756 middle and high school students, which was a 23% increase from last year, and 58% increase from the year 2001. Contestants came from all over Colorado: Aurora, Berthoud, Buena Vista, Calhan, Centennial, Colorado Springs, Denver, Dinosaur, Falcon, Firestone, Fort Collins, Fountain, Franktown, Frederick, Lakewood, Loveland, Manitou Springs, Monument, Parker, Peyton, Pueblo, Rangely, Twin Lakes, Widefield, and Woodland Park.

The judges awarded two first prizes: to Bryce Herdt, now a sophomore from Mitchell High School, and to Mark Heim, a sophomore from Thompson Valley High School, who participated for the first time. Each of them received the gold medal of the Olympiad, the CASIO *Class Pad* software, the software *Explorations* donated by Wolfram Research from Silicon Valley, a Voyager 200 donated by Texas Instruments, the book *A New Kind of Science* written by the founder of Wolfram Research, Inc., Dr. Steven Wolfram, and a $2,000 scholarship to be used at any accredited American university or four-year college.

This was third victory for Bryce, who was the youngest winner ever in 2001, and the first victory for Mark. Do not forget Mark – for he will blow the Olympiad field away in the next two years, before sailing for M.I.T.!

Second prize was awarded to Steven Biles, a senior from Palmer High School, and a last year winner. He received the silver medal of the Olympiad, the CASIO *Class Pad* software, the Texas Instruments' Voyager 200, the book *A New Kind of Science* by Steven Wolfram, a $1,000 scholarship, and the book *Geometric Etudes in Combinatorial Mathematics* by V. Boltyanski and A. Soifer.

Third prizes were presented to the following two Olympians: Helen Saltthorpe, a junior from Arapahoe High School; and Christopher Liebmann, a sophomore from Wasson High School. Each of them received a CASIO graphing calculator, the CASIO Classpad software and Steven Wolfram's book.

As you recall, UCCS Chancellor Pamela Shockley-Zalaback instituted a "Chancellor's Scholarship for CMO Medalists" effective with last year's Olympiad. Each gold, silver and bronze medalist had been awarded a $1000 scholarship, effective upon acceptance to UCCS as a new freshman. This year the Chancellor's offer amounted to up to $5,000 since we had medalists.

This year the judges also awarded 2 fourth prizes, 15 first honorable mentions, and 108 second honorable mentions. All 130 winners received Wolfram Research's mathematical T-shirts. Sizes were subject to availability, or last placed first choose :-).

The Literary Award was presented to Brent Seward, a senior from Palmer High School for

Ode to Problem 20.2 *Quarreling Jacques*

Quarreling Jacques moves from block to block
He can't seem to make any friends
Get the hell out they scream and they shout
The location he moves just depends

I'll hop in my car and go as far
As far as I can in Paris
But when my children and I are tired of moving
The neighbors will just have to bear us.

Two of the Olympiad's longest-serving judges. From the left, Shane Holloway and Jerry Klemm.

The Prize Fund of The Olympiad was generously donated by Ingersoll-Rand Company, Wolfram Research, Inc., CASIO, Texas Instruments, Air Academy School District 20, Harrison School District 2, Colorado Springs School District 11, Bennett School District 29J, St. Mary's School, Sand Creek High School, Mann

Middle School, Arapahoe High School, Fort Collins High School, Rangely School District RE-4, The Colorado College, UCCS Vice-Chancellor's of Academic Affairs, and UCCS Bookstore.

The Award Presentation Program included *Review of Solutions of the Olympiad Problems* by Alexander Soifer and his lecture *Etudes on Coloring*.

The following guests of honor, hosts, and sponsors addressed the winners and present the awards: Pamela Shockley-Zalaback, Chancellor of CU-Colorado Springs; Michael Merrifield, House Member of the Colorado State Congress; Lionel Rivera, City of Colorado Springs; Norman Ritter, Superintendent of Colorado Springs School District 11; and Gregory C. Hoffman, Director, Human Resources, Ingersoll-Rand Company.

In the 20 years of the Colorado Mathematical Olympiad, 14,000 students have participated during 1984-2003. They wrote some 69,000 essays, and were awarded some $171,000 in prizes. The Olympiad has been a unique joint effort of school districts, schools, institutions of higher education, the business community, and local and State governments. It is the largest essay-type mathematics competition in the USA.

Everything looked sunny for the Olympiad, as it should: in Colorado we have 300 sunny days a year. Then in July 2002 came a disaster, named Linda Nolan, the new Dean of the College of Letters, Arts and Sciences (LAS).

At the end of *Historical Notes* 8, you read *"Geombinatorics* Is Born." By now *Geombinatorics* has served high school, college, and professional mathematicians for a dozen years. Imagine, one of the first orders of the new dean was to forbid any and all clerical support for *Geombinatorics*! I replied to Dean Nolan on July 25, 2002:

> I am amazed – I truly am – that a brand new dean on her day one would try to murder a 12-year old, teenage journal, which provided students and professionals with open problems and helping them to enter research. But I shouldn't really be surprised, for the notorious Dean Tracey similarly ordered me to close down the Colorado Math Olympiad when she was a 2-year old baby in 1986.[1]

---

[1] You have read about Dean Tracey's affair in *Historical Notes* 3.

Apparently, Dean Nolan decided to remove the only defender of *Geombinatorics*. The University Gallery Director Gerry Riggs attended a meeting where the dean stated: "We are building a file and just waiting for him [Soifer] to step over the line."

I felt that only *glasnost* could help us, and formed "The Academic Rights Watch Group" (ARWG). On September 22, 2002, I wrote about it to all faculty and staff of my campus:

> *Silence is golden.*
> An old proverb

In a recent nightmare I saw blood flowing from the LAS Dean's office down the long hall (remember Kubrick's "Shining"?)... And I saw those who claimed, "no, milk and honey were flowing there," and immediately formed a line for little perks...

Seriously, we, lowly employees, do not have much protection. Any time a faculty dares to file an internal grievance against, say, a dean, the faculty finds herself alone against the system, as the dean immediately gets a University lawyers and patronage of a Vice Chancellor and higher ups. And when a faculty files with the State or the Feds, a year later the faculty could get at best a "permission to sue"...

The staff's position is even worse. I personally know staff members who were told by their UCCS bosses to tone down their rhetoric or better yet stop participating in e-mail discussions altogether. And even when staff members win against oppressive bosses at the State Personnel Board, the University's deep pockets and boss's petty vindictiveness would likely strip the staff of their victory on appeal... I hear all the time that serving at a boss's pleasure and standing to be fired at will, reminds employees the old proverb: *Silence is golden.*

Worse yet: you, the grievant, are told that for *your* protection all relevant papers get this notorious stamp *"Confidential Personnel Matter"* and are hidden from all-liberating light of glasnost. Consequently, the grievant never gets to know that there are other oppressed employees. So, the confidentiality protects the oppressor much more than it protects the oppressed. The boss gets all the ammo of the system to fight each of us separately, the entire system against one lowly employee. I know, I've been there... And even though I've won, it wasn't easy nor fun :-(.

My favorite Supreme Court Justice, the late Harry Andrew Blackmun (the author of the January 22, 1973, Court's Opinion in 410 U.S. 113 "Roe, et al. v. Wade") understood that abuse of confidentiality by the administrators very well, when on January 9, 1990, he wrote for the unanimous U.S. Supreme Court (University of Pennsylvania v. EEOC, 493 U.S. 182, 1990) *"Indeed, if there is a "smoking gun" to be found that demonstrates discrimination in tenure decisions, it is likely to be tucked away in peer review files."*

This is why I am writing to you today: we ought to get together and extinguish smoking guns. We need glasnost here, we ought to light up dark corners of secret personnel files – especially now, when it seems that our boss is dead bent on creating KGB-type files specifically for combating her own employees.

I hereby announce the creation of Academic Rights Watch Group, ARWG for short. It would work on our small scale like its famous counterpart did on the world stage: the Helsinki Human Rights Watch Group was effective in exposing and limiting the extent of the Soviet ruthless violations of human rights by putting the latter in the bright light of public view.

ARWG would offer staff & faculty grievants a sympathetic ear and valuable advice. ARWG would offer a moral and practical support to individuals believed to be abused by an administrator and the system.

ARWG would compile violations to see the whole picture, to learn the state of the college and to go to higher University and State authorities for urgent help when needed. With permission of grievants, ARWG would bring publicity to violations, and this publicity would likely affect the oppressors in a way only glasnost can.

And while I see the present urgency in offering the ARWG services to LAS employees, I am sure we would find time and energy to help others on campus.

I invite "the few, the proud, the marines" to join me in forming ARWG.

I invite my colleagues – staff & faculty in need of help – to contact me by phone (listed :-), e-mail, or simply drop by my office. Yes, I am leaving for Princeton in three months – but that is all right, as I would continue with ARWG wherever I am.

Do not forget the warning of the late Great Russian poet and dissident Alexander Galich:

Silence is golden,
Be silent – and you will join the rich,

Be silent – and you will join the executioners,
Be silent, Be silent, Be silent.

I formed ARWG with four more faculty and staff members, totaling 120+ years of the university service. Even though ARWG was a grass-roots movement, the chancellor recognized ARWG and agreed to meet and work with us.

It took Dean Nolan a while, but in 2003 she attempted to close down the Colorado Mathematical Olympiad as well. I was 2000 miles away from Colorado as a Visiting Fellow at Princeton University when the dean's letter of August 19, 2003 reached me in Princeton on September 2:

> Dear Alex,
>
> We hope that you are enjoying your sabbatical leave. As you are aware, UCCS has been struggling with a severe state budget cut... Using guidelines given to us by the Chancellor, we had to cut our base by 1.5%, which is a permanent reduction.
>
> Unfortunately, the $7,400 that the College of LAS funds towards the support of the Math Olympiad has fallen victim to the state budget cuts....
>
> I sincerely regret that the College of LAS cannot continue monetary support for the very successful Math Olympiad at this time.

Did you get it? The college has to cut 1.5%, and the Olympiad is cut its fair share of 100% of the budget! I must have not slept very well, for the next morning at 5 A.M. I sent my e-mail comments to UCCS Chancellor Shockley Zalaback:

> The Olympiad has been an enormous undertaking that produced immeasurable benefits. Any normal place would celebrate my 20 years of fanatically dedicated work on the Olympiad by valuable awards and medals. UCCS is no normal place; it is a circus: I have managed to run the Olympiad on a shoestring budget, and now Nolan asked me to perform a new trick of running it without any budget!

The very next day, I was assured of the overwhelming support of the chancellor's office. First Kathy Griffith, Executive Assistant to the Chancellor, replied to the dean:

Linda,

I need to inform of you [of] the following: The Math Olympiad account was established in 1987-88 by contract between Alex Soifer and the University. The account is only housed in the College of Letters, Arts and Sciences... At any rate, you will be unable to use the $7,400 toward your base budget cut of 1.5%. I will be requesting soon that this account be transferred to the Chancellor's Office since the Math Olympiad is temporarily being managed from our office. Thank you.

Then Chancellor Pamela Shockley sent me her own assurance:

You must know that I want the Olympiad to continue and have every intention that it will do so.

I remained in Princeton, but these messages from Kathy and Pam restored in my mind's eye the sunny, blue-skied Colorado day!

Problems 20

Problems 20.1 and 20.3 were created and contributed to us by Alexei Kanel-Belov of Moscow, who at the time of this Olympiad was a Professor at Bremen, Germany, and later moved to Israel. A. Soifer was inspired by the Russian mathematical folklore to create Problem 20.2. Problem 20.4 was contributed by Dr. Ming Song of Colorado Springs. Problem 20.5 was created by Alexander Kovaldzhi of Moscow. These problems were selected and edited for you by the Problem Committee: Robert Ewell and Alexander Soifer.

20.1 COUNTING TUESDAYS (*A. Kanel-Belov*). The numbers of Mondays and Wednesdays in the year Y are equal. Must the number of Tuesdays be equal to them if

(A) Y is a regular year;
(B) Y is a leap year?

20.2 QUARRELING JACQUES[2] (*A. Soifer*). Jacques lives at a point A_1 of Paris (Paris forms a geometric figure in the plane). He

---

[2] This problem is my homage to Jacques Chirac, President of France 1995-2007.

quarrels with everyone, and moves to a point A_2 as far away from A_1 as possible in Paris. He quarrels there too, and moves to point A_3 as far away as possible from A_2. Jacques continues quarreling, always moving to a point farthest away from where he is. Prove that if A_3 must differ from A_1, then Jacques will never move back to A_1.

20.3 MATH 101 (*A. Kanel-Belov*). Is there a way to place all positive integers $1, 2, \ldots, n, \ldots$, one integer per cell, filling all cells of an infinite square grid, so that the sum of any 10 consecutive numbers in a row or column is divisible by 101?

20.4 20 COINS FOR 20 YEARS (*Ming Song*). There are 20 coins of equal size on a rectangular sheet of paper with the centers of all coins on the paper and without any overlap between coins. If no more coins of the same size can be put onto the paper with the coin's center on the paper and without overlap with any one of the 20 original coins, prove that we can use 80 coins of the same size to cover the sheet of paper. (Note: *Covering* the sheet of paper means that you cannot see any point of the sheet of paper from the top view; coins have 0 thicknesses; of course, some coins among the 80 coins must overlap.)

20.5 A MAP COLORING GAME (*A. Kovaldzhi*).

(A) The explorer and the mapmaker are taking turns in a map coloring game. At every turn the explorer draws a new contiguous country in the plane with no inside points in common with the previously drawn countries. The mapmaker then colors the new country so that no two countries of the same color share a boundary line. (They are allowed to share one or even finitely many points.) The explorer wins if he forces the mapmaker to use at least 5 colors; otherwise the mapmaker wins. Find a strategy that allows one player to win regardless of how the other plays. The explorer, naturally, goes first.
(B) Solve the same problem with 2003 colors replacing 5 colors.

Solutions 20

20.1. Let us introduce a convenient shorthand. If a gives remainder r upon division by n, we will record it as $a \equiv r \pmod{n}$. For example,

$$365 \equiv 1 \,(\mathrm{mod}\,7) \tag{1}$$

$$366 \equiv 2 \,(\mathrm{mod}\,7) \tag{2}$$

But (1) means that we have equal numbers of six days of the week, while one day of the week may occur one extra time – this could be Tuesday! Thus for a regular year the answer is no.

On the other hand, (2) means that there are *two consecutive* days of the week that have one extra day, compared to five other days of the week. Since Monday and Wednesday are not consecutive, they must belong to the "five other days of the week." These five days are consecutive as well, and so Tuesday belongs to them. Thus, for a leap year the answer is yes. ■

20.2.

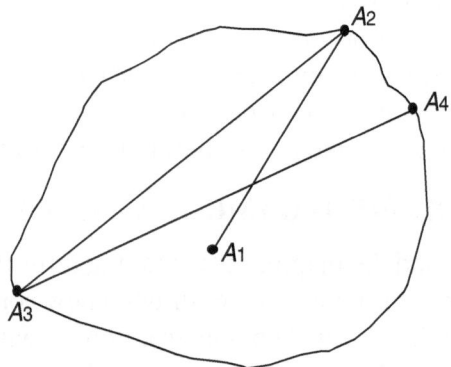

Fig. 20.1

Let $|AB|$ denote the length of a segment AB. Observe that $|A_{n-1}A_n| \leq |A_n A_{n+1}|$, for we choose A_{n+1} to be a farthest from A_n (Figure 20.1), and therefore A_{n+1} is at least as far from A_n as A_{n-1} is. So, we have a sequence

$$|A_1 A_2| < |A_2 A_3| \leq |A_3 A_4| \leq \ldots \leq |A_n A_{n+1}|. \tag{3}$$

Assume now that Jacques gets back to A_1; this means that $A_{n+1} = A_1$ for some n, which in view of (3) implies $|A_1 A_2| < |A_n A_1|$. The latter inequality contradicts the choice of A_2 as the farthest point from A_1.

Remark: Bryce Herdt put the same argument quite concisely:

"Suppose Jacques can return to A_1. Then all the moves he makes will form a loop. Since consecutive distances can't decrease, they can't increase either and must all be the same. But then it's not the case that A_3 must, as stated, differ from A_1. Therefore, Jacques won't return to A_1." ■

20.3.

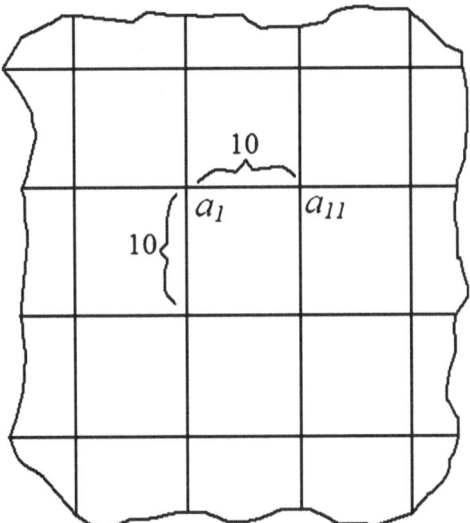

Fig. 20.2

The shorthand introduced in the solution of Problem 20.1 above can be expanded as follows: if a and b give the same remainder upon division by n, we record it as $a \equiv b \pmod{n}$.

Assume we have placed all integers $1, 2, \ldots, n, \ldots$, one integer per cell, filling all cells of an infinite square grid, so that the sum of any 10 consecutive numbers in a row or column is divisible by 101. Look (Figure 20.2) at any consecutive 11 numbers in a row or column: $a_1, a_2, \ldots, a_{10}, a_{11}$. We observe that

$$a_1 + a_2 + \ldots + a_{10} \equiv a_2 + a_3 + \ldots + a_{10} + a_{11} \pmod{101};$$

Therefore, $a_1 \equiv a_{11} \pmod{101}$. So, we have established that the remainders upon division by 101 periodically repeat. This implies that all remainders present in the infinite grid appear already in a 10×10 square S of the grid. Indeed, the infinite grid can be tiled by the vertical and horizontal translates of S, and periodicity guarantees the repetition of the same remainders. Thus, we have at most $10 \times 10 = 100$ different remainders upon division by 101 represented among the numbers on the infinite grid. But surely, there are 101 distinct remainders. This contradiction proves that the required placement does not exist. ∎

20.4. Assume that the radius of the coins is 1. Let A be any point on the sheet. The distance between A and the center of the coin closest to A must be less than 2, for otherwise an additional coin could be placed on the sheet with A as its center (Fig. 20.3).

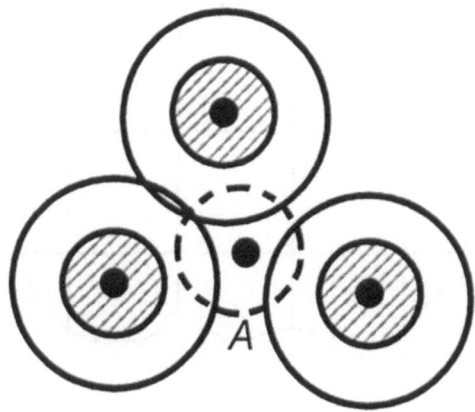

Fig. 20.3

Therefore, if we doubled the diameters of each of the 20 coins (and kept their centers intact), the 20 coins would completely cover the sheet. Denote by C this covering of the given sheet by 20 coins of double diameter.

A homothetic image $C_{1/2}$ of C with coefficient $1/2$ (i.e., shrinking of all linear sizes of C by a factor of 2, including shrinking of the radius of the coins) would give us a complete covering of a sheet half the given sheet length and width by 20 coins of original size. Now we can cover the given sheet by 80 given coins: just partition the given sheet into four congruent parts (Fig. 20.4), and cover each part by the covering $C_{1/2}$.

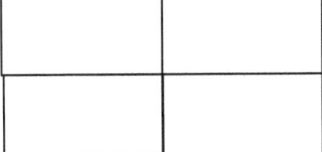

Fig. 20.4

Observe that the two operations that have been used in the solution, are commutative: one can first shrink the original sheet with 20 coins on it by a factor of 2, and then cover it by double size coins (i.e., coins of the originally given size), and finally put four such sheets together as in Figure 20.4 In fact, the only solution in the Olympiad was obtained by Mark Heim, who used this order of operations. ∎

20.5. (A). A map M can be represented by a graph $G(M)$ as follows: choose a vertex (the capital city) inside each country – these points form the set of vertices of $G(M)$. We connect two vertices by a line, or "edge", if the corresponding countries share a boundary line – thus we form the set of edges of $G(M)$, and our graph is created. Please observe that we can draw $G(M)$ without intersection of its edges by drawing the edges through common boundaries of the countries. A graph that can be drawn without intersection of its edges is called *planar*. A graph that *is* drawn without intersection of its edges is called *plane*.

Our Map Coloring Game has therefore been translated into the language of graphs: in each move the explorer draws a new vertex and attaches it by edges to some of the existing vertices without intersection of the edges. The mapmaker then colors the new vertex in such a way that it is not connected by an edge to a vertex of the same color.

Solution of Problem 5(A). By his first four turns the explorer creates what we call K_4, the graph with four pairwise connected vertices (for example, the three vertices of an equilateral triangle and its center, with six connecting lines). The mapmaker is forced to use four different colors on these four vertices (Figure 20.5).

We have three colors on the "outside," say, colors 2, 3, and 4. The explorer now draws a point P without any edges. If the mapmaker colors P in any color n that is not in the color set $\{2, 3, 4\}$, we are ready for "the final assault."

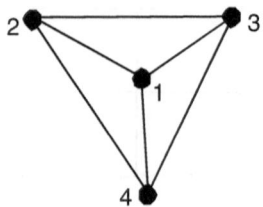

Fig. 20.5

If the explorer colors P in one of the colors 2, 3, or 4 – say, 2 – then in his next three turns the explorer surrounds P by a triangle to form another K_4 component with P in the middle, thus forcing the mapmaker to use color n that is not in the color set $\{2, 3, 4\}$ on one of the three added points (n could be 1), and we are ready for "the final assault."

In either of the above two cases, in his "final assault" the explorer draws a new vertex Q and connects it to vertices of colors 2, 3, 4 and n, thus forcing the mapmaker to use the fifth color on Q (Figure 20.6 and Figure 20.7), and we are done. ■

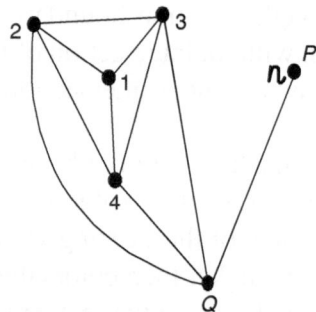

Fig. 20.6

20.5. (B). Surely, we can push the above type of construction further and easily force 6 colors, 7, 8. But forcing 2003 colors requires a "system." And the cycles, or closed walks created in the solution of Problem 20.5(A), are not helpful. This suggests an idea to use no cycles in the explorer's construction. And "the final assault" gives the necessary building idea, an "inductive step."

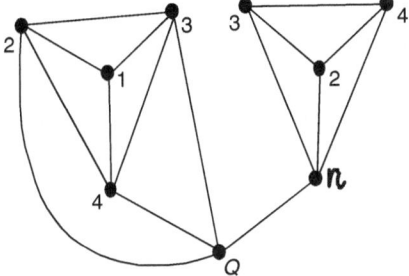

Fig. 20.7

A graph without cycles (i.e., without closed walks along the edges) is called *forest*. Very appropriately, a forest consists of connected components, called *trees*.

We will prove by induction that for any positive integer n, the explorer can construct a forest F containing n trees F_i, $i = 1, 2, \ldots, n$, such that we can choose one vertex v_i per tree F_i so that all vertices v_i, $i = 1, 2, \ldots, n$, have been colored in different colors by the mapmaker.

There is not much to do for $n = 1$. The explorer draws one point, and we are done.

Assume that the statement is true for $n = k$, i.e., the explorer can construct a forest F containing k trees F_i, $i = 1, 2, \ldots, k$, such that we can choose one vertex v_i per tree F_i so that all vertices v_i, $i = 1, 2, \ldots, k$, have been colored in different colors by the mapmaker.

Let now $n = k + 1$. We first let the explorer exercise the inductive assumption and construct a forest F containing k trees F_i, $i = 1, 2, \ldots, k$, such that we can choose one vertex v_i per tree F_i so that all vertices v_i, $i = 1, 2, \ldots, k$, have been colored in different colors by the mapmaker (see the left side of Figure 20.8).

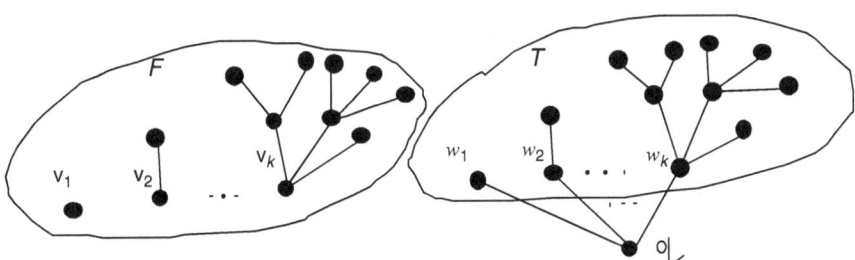

Fig. 20.8

We can then let the explorer repeat the same construction again, of a forest T containing k trees T_i, $i = 1, 2, \ldots, k$, such that we can choose one vertex w_i per tree T_i so that all vertices w_i, $i = 1, 2, \ldots, k$ have been colored in different colors by the mapmaker.

If the forest T has a vertex v_{k+1} colored in a color not represented among the colors of the k vertices v_i, $i = 1, 2, \ldots, k$, then we add the tree containing v_{k+1} to the forest F to obtain the required forest.

If the mapmaker played "smartly" and colored the forest T without using colors other than those of the k vertices v_i, $i = 1, 2, \ldots, k$, then the explorer can draw a new vertex q and connect it to each of the vertices w_i, $i = 1, 2, \ldots, k$, thus creating a tree that we will denote by F_{k+1}. The mapmaker now must color q in a color not represented among the vertices w_i, $i = 1, 2, \ldots, k$. The required forest has now been constructed: it is F_i, $i = 1, 2, \ldots, k, k + 1$ (Figure 20.8). The induction is complete.

If the mapmaker played "smartly" from the beginning and colored the forest T without using new colors unless necessary, the early forest would look like the one shown in Figure 20.9, in which numbers stand for colors: number "4" denotes color 4.

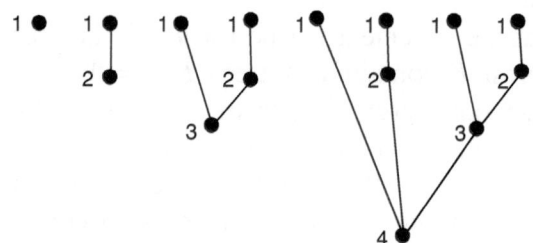

Fig. 20.9

This problem was essentially solved by two contestants, Bryce Herdt and Steven Biles.

We will continue map-coloring explorations in our final *Further Exploration* E20. ■

The Story After

Having returned to Princeton after the Olympiad, on Monday, April 28, I was having lunch at the Institute for Advanced Study with members of the *Computer Science/Discrete Math Seminar*, following the talk "Partial Results on the Total Colouring Conjecture" by the well-known graph theorist Bruce Reed from McGill University, Montreal, Canada. Princeton University's Paul Seymour sat across the table from me. Paul is one of the world's great graph theorists and a co-author of the second, better proof of the Four-Color Theorem. I felt compelled to test the graph celebrities on the problems of our Olympiad!

"How many colors suffice to color a tree?" I asked Paul.

"You must be kidding," was his reply.

"I am, but here is the problem," and I formulated the game of explorer and mapmaker of Problem 20.5, and asked "can the explorer force any number of colors?"

Within seconds both Paul and Bruce independently solved the problem. I was "in troubles," and so I offered them "20 Coins for 20 Years," Problem 20.4.

Paul quickly observed that double size coins would completely cover the sheet. But the progress lost steam. Bruce, Paul, a few young gifted Ph.D.'s and Ph.D. students worked fast and hard, with no result. I was asked by Paul to show the coverage by four coins when just one coin was given, which I did without a problem. My offer to show a complete solution was rejected, and the problem was left as "homework."

Two days later a perfect solution was found by a Princeton's third year Ph.D. student from the U.K. Peter Keevash. (Since then, Peter defended his doctorate at Princeton and after a stint at Caltech, returned to his homeland, where he is a Lecturer in Pure Mathematics at the School of Mathematical Sciences, Queen Mary, University of London.)

I also gave the Coins problem, with a solution, at lunch to a group of veterans of the Princeton Math and Physics Departments in the historic Prospect House in Princeton, including some members of the National Academy of Sciences and Nobel Laureates. "This is very beautiful," responded Edward Nelson, a Princeton Professor since the 1950s and the creator of the great open *chromatic number of the plane problem*, which we have discussed in *Further Exploration* E2.

By the way, the chromatic number problem is still open, and thus I am tempted to assign you the following homework: narrow down the range of possible answers to this problem from the present 4, 5, 6, or 7 – and let me know!

Part IV
Further Explorations of The Second Decade

The teaching of mathematics has sometimes degenerated into empty drill in problem solving, which may develop formal ability but does not lead to real understanding or to greater intellectual independence...

Understanding of mathematics cannot be transmitted by painless entertainment any more than education in music can be brought by the most brilliant journalism to those who have never listened intensively. Actual contact with the content of living mathematics is necessary.

– Richard Courant
What is Mathematics? Preface to the 1st ed., 1941

How I Envision Mathematics Education: Introduction to Part IV

When I was in middle school – in grades 6, 7, 8 – my first Olympiads showed that there is mathematics unlike the one in school, beautiful, humorous, surprising. Yet in retrospect, I have to admit that even the Olympiads did not give me answers to such fundamental questions as

(1) *what is mathematics?*
(2) *what do mathematicians do?*

This ought to change. Math instruction in school ought to answer these questions, as history instruction ought to show what history is and what historians do. How shall we do it?

We all agree that *problem solving* is the most important component of math and math learning. We disagree on what problem solving should amount to. It is mostly used as a synonym of *analysis*, the search for a solution of a *single-idea* problem. A typical problem might be: "given legs 3 and 4 of a right triangle, compute the hypotenuse by using the Pythagorean Theorem." (Now, in 2010, due to alleged "progress," we hear "given legs 3.1 and 4.2 of a right triangle, compute the hypotenuse using the Pythagorean Theorem *and your calculator*.") This does not even begin to show a student *what mathematics is and what mathematicians do*.

We ought to use *multi-idea* problems, requiring for their solution a *synthesis* of ideas and methods from different areas of mathematics. We ought to introduce students not only to analytical reasoning but also to *construction* of counterexamples. As I. M. Gel'fand said,

Theories come and go; examples stay forever.

In order to show what mathematics is and what mathematicians do, we ought to present *real fragments of mathematics* with their analytical proofs and constructions of counterexamples; with *open-ended*

A. Soifer, *The Colorado Mathematical Olympiad and Further Explorations:* 323
From the Mountains of Colorado to the Peaks of Mathematics,
DOI 10.1007/978-0-387-75472-7_35, © Alexander Soifer, 2011

and open problems; with *mathematical intuition* leading research like a light at the end of a tunnel; with synthesis of ideas from algebra, geometry, trigonometry, linear algebra, mathematical analysis, combinatorial geometry.

If we are to succeed in passing the baton to following generations, we must show mathematics with its *beauty, elegance, and results challenging our intuition.* We ought to try to show that *mathematics is alive,* that every solved problem gives birth to a myriad of unsolved problems. We want to *stop discrimination based on age,* and offer young talented mathematicians who are still in high school or early college the opportunity to be the first to solve a problem. As the *South Carolina Reflector* put it,

> Think like a tea bag. You don't know your strength until you get in hot water.

Mathematics cannot be taught; it can only be learned by students doing math with our gentle guidance. As Courant wrote in 1941, *"Actual contact with the content of living mathematics is necessary."*

Looking back, it seems that I have dedicated my life to showing young mathematicians what mathematics is, through all my activities: my 7 books and most of my 200+ articles, 28 years of the Colorado Mathematical Olympiad, and 20 years of the research quarterly *Geombinatorics* dedicated to problem-posing essays that allow young mathematicians to engage in real research. All of them present "live mathematics", all of them are written – I hope – in the inviting style of a dialog. All of them create *bridges between problems of mathematical Olympiads and research problems of mathematics.* And the influence is mutual: research is an invaluable source of new ideas for Olympiad problems, and Olympiad problems inspire new investigations!

I welcome you to ten new such bridges, *Further Explorations of the Second Decade!*

E11. Chromatic Number of a Grid

Inspired by problem 11.4

As you recall, Problem 11.4 was created by Professor John Horton Conway of Princeton University and me on March 8, 1994, when we met at the "International Southeastern Conference on Combinatorics, Graph Theory, and Computing" at Florida Atlantic University in Boca Raton, Florida. I had heard a lot about John, and read his (jointly with Elwyn Berlecamp and Richard Guy) spectacular book *Winning Ways for Your Mathematical Plays*. John's talk about quantum mechanics was very entertaining (at one point he even jumped up in the air :-) and inspiring.

I mentioned to John my favorite unsolved problem of finding the chromatic number of the plane (see *Further Exploration* E2), and this prompted us to define and calculate the chromatic number of a square grid, which became Problem 11.4. We then moved on to a triangular grid. Before you follow us, we ought to formulate the definition of the *chromatic number of a grid* in general.

Given a grid G and a positive number d, let $\chi(d)$ be the minimum number of colors required for coloring the vertices of the grid G so that any two vertices a distance d apart are colored in different colors. The maximum χ of $\chi(d)$ over all d is called the *chromatic number of the grid G*.

Can you find the chromatic number of the triangular grid (Figure E11.1)?

A. Soifer, *The Colorado Mathematical Olympiad and Further Explorations: From the Mountains of Colorado to the Peaks of Mathematics*, DOI 10.1007/978-0-387-75472-7_36, © Alexander Soifer, 2011

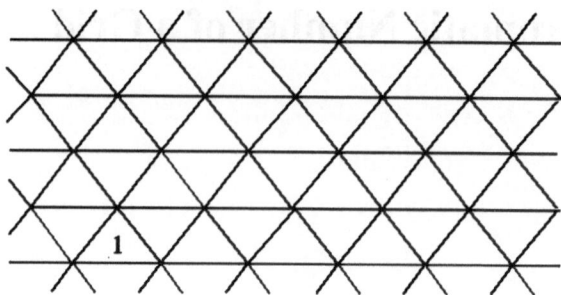

Fig. E11.1

E12. Stone Age Entertainment

Inspired by problem 14.1

In Problem 14.1, the game allowed a nice solution for the pile of 99 pebbles. What would happen if the pile had, say, 100 pebbles? Imagine, a different player would win! Once again we can exclaim, behold:

$$1 + 3 + 5 + \ldots + 19 = 100.$$

And once again this immediately gives a winning strategy, but now for Fred Flintstone: he takes one pebble, and on each subsequent move Fred takes the difference between 3, 5, …, 19 and the number of pebbles taken by Barney Rubble respectively. Fred's strategy allows for a nice geometric illustration, so nice that I would call it a *proof without words*, a term used before in math, even on book titles, and must have been – or should have been – inspired by Felix Mendelssohn's *Songs without Words* (Figure E12.1).

On his first move Fred takes the top left square of the $n \times n$ grid. Fred then puts the number of pebbles taken by Barney into consecutive L-shaped layers and takes all the remaining pebbles from that "L".

OK, now we know how to play with 99- and 100-pebble piles. But who has a winning strategy if the original pile consists of, say, 2010 pebbles? (It is customary to ask Olympians to solve a problem for the year we live in, for it usually provides a large enough, arbitrary enough number, so that to solve for it is virtually equivalent to solving for all n. I am betting here that this book will leave the Springer birth center in 2010 :-).

First of all, we can readily generalize the result from 100 to any perfect square number of pebbles, for in general we have the following equality:

A. Soifer, *The Colorado Mathematical Olympiad and Further Explorations:*
From the Mountains of Colorado to the Peaks of Mathematics,
DOI 10.1007/978-0-387-75472-7_37, © Alexander Soifer, 2011

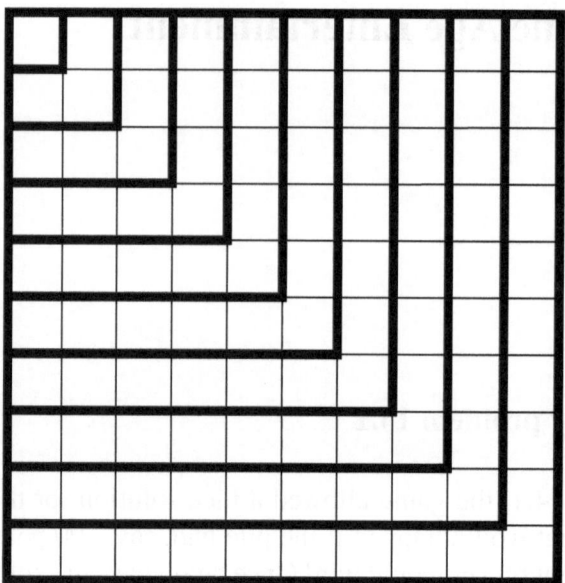

Fig. E12.1

$$1 + 3 + 5 + \ldots + (2n - 1) = n^2.$$

Thus, for a pile of n^2 pebbles Fred wins by taking one pebble on the first move, and then taking the compliment of Barney's draw to $3, 5, \ldots, (2n - 1)$ respectively. But what can one do if the number of pebbles in the pile is not a perfect square?

It is critical to notice that *on his second move Fred is not limited to taking the complement of Barney's draw to 3 – Fred can take the complement to 4 as well!* On his next move Fred can take the complement of Barney's draw to 5 or to 6, etc. (Figure E12.2).

Thus, after his n moves Fred can make sure that instead of n^2, as many as $n^2 + n - 1$ pebbles are removed from the pile, for we have the equality

$$1 + 4 + 6 + \ldots + 2n = 2\,(1 + 2 + 3 + \ldots + n) - 1 = n^2 + n - 1.$$

But of course, Fred can choose how many times he takes the complement to a member of the odd number sequence $3, 5, \ldots, (2n - 1)$ and how many times Fred takes the complement to a member of the even

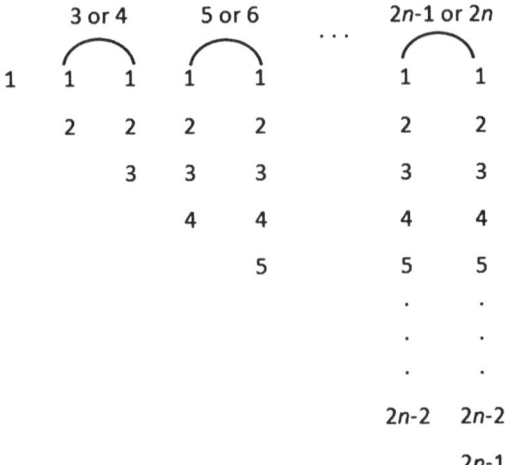

Fig. E12.2

number sequence $4, \ldots, 2n$. Thus, Fred has a winning strategy for any number of pebbles in the original pile between n^2 and $n^2 + n - 1$.

Fred's winning strategies can be nicely illustrated geometrically (Figure E12.3).

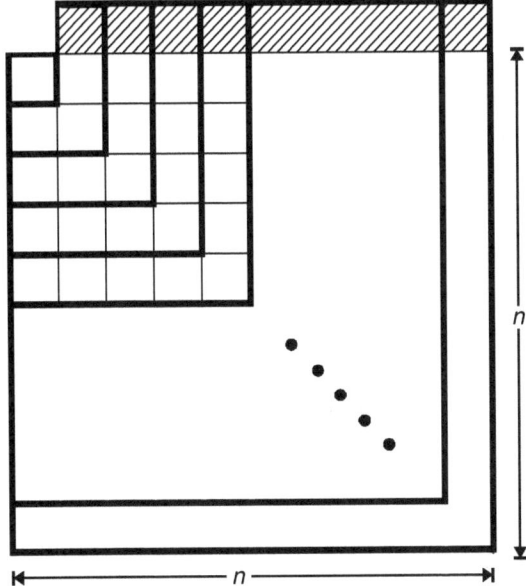

Fig. E12.3

But what can we do if the number of pebbles in the original pile is in the range $n^2 + n$ to $n^2 + 2n$?

Well, let us look at the game through Barney's clever eyes! On his first move, Barney can take the complement of Fred's first draw (Fred must draw 1 pebble) to 2 or to 3. On his next move, Barney can take the complement of Fred's draw to 3 or to 4, etc. (Figure E12.4).

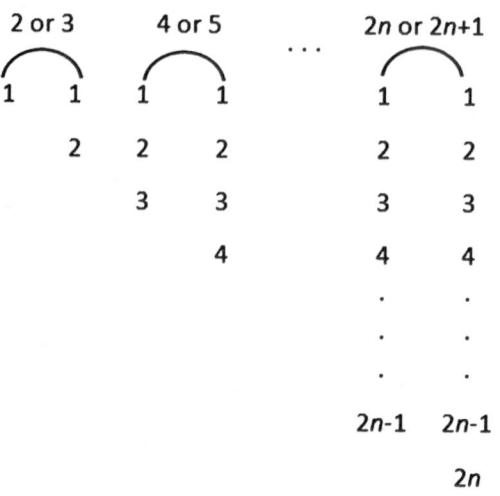

Fig. E12.4

Observe the following equalities and prove them on your own:

$$2 + 4 + 6 + \ldots + 2n = n^2 + n;$$
$$3 + 5 + 7 + \ldots + (2n + 1) = n^2 + 2n.$$

Of course, Barney can choose how many times he takes the complement to a member of the even sequence $2, 4, 6 \ldots, 2n$ and how many times he takes the complement to a member of the odd sequence $3, 5, 7 \ldots, (2n + 1)$. Thus, Barney has a winning strategy for any number of pebbles in the original pile between $n^2 + n$ and $n^2 + 2n$.

Now we have the criterion for determining who wins the game with a pile of m pebbles before the game starts. First we find the largest perfect square n^2 that does not exceed m. If $n^2 \leq m \leq n^2 + n - 1$, then Fred wins. If $n^2 + n \leq m \leq n^2 + 2n$, then Barney wins. Figure E12.5 shows at once this criterion and strategies for the winner.

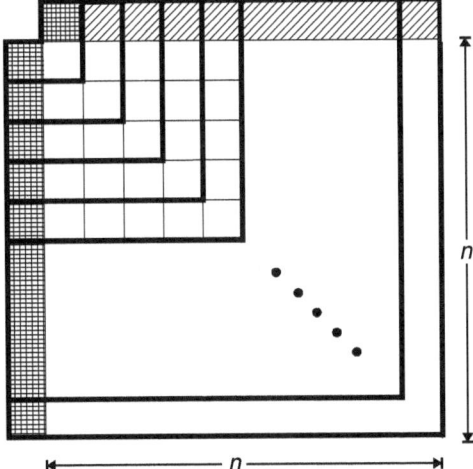

Fig. E12.5

If m can be assembled of n^2 squares of the $n \times n$ grid plus a number of striped squares on top, i.e., $m = n^2 + k$, where $0 < k \leq n - 1$, Fred wins. On his first move Fred takes the top left square of the $n \times n$ grid. Fred then puts the number of pebbles taken by Barney into consecutive L-shaped layers and takes all remaining pebbles from that L plus one striped square above the L as many times as necessary, namely k times, to assemble m; and he ignores all checkered squares.

If m can be assembled of n^2 squares of the $n \times n$ grid plus *all* striped squares at the top, plus one checkered square on top, plus possibly a number of checkered squares on the left side, i.e., $m = n^2 + n + k$, where $0 \leq k \leq n$, then Barney wins. Barney puts the number of pebbles taken by Fred into consecutive L-shaped layers and takes all remaining pebbles from that L, plus one striped square above the L, plus one checkered square on the left side as many times as necessary, namely k times, to assemble m.

In particular, $2010 = 44^2 + 44 + 30$, and so Barney has a winning strategy in the case of the 2010-pebble pile, and we are done!

E13. The Erdős Train Station

Inspired by problem 14.4

Paul Erdős's articles remind me of train stations. They do not merely report a result or two that the author proved. Each Erdős's classic papers is a treasure trove of many trains of thought. Let us take a ride here on a couple of such trains from the Erdős Station-1946 [E0].

In this paper Paul actually proves more than I asked in Problems 14.4(A) and (B). We will first ride Train A.

Train A

Minimal Distance Result E13.1. (P. Erdős, 1946). In any set of n points in the plane the minimum distance r can occur at most $3n - 6$ times.

Proof Two minimum distance segments cannot intersect, for otherwise they would generate a shorter segment (prove this nice geometric fact on your own) in contradiction to their minimality.

Since the minimum distance segments do not intersect, they (together with their endpoints of course) form a plane graph, and by the Euler Formula we get the upper bound of $3n - 6$ for the number of edges in a plane graph. For the Euler Formula and this corollary see, for example, Section 11, "Planarity," in [S14] or [BS]. ∎

Paul Erdős does not stop at this result. At the end of the paper, he writes:

By more complicated arguments we can prove that the minimal distance r' can occur not more than $3n - cn^{1/2}$ times, where c is

A. Soifer, *The Colorado Mathematical Olympiad and Further Explorations: From the Mountains of Colorado to the Peaks of Mathematics*, DOI 10.1007/978-0-387-75472-7_38, © Alexander Soifer, 2011

a constant. On the other hand the example of the triangular lattice shows that r' can occur not more than $3n - cn^{1/2}$ times. I did not succeed in determining exactly how often r' can occur.

It is nothing short of amazing to see how great Paul Erdős's intuition was. All these Erdős' conjectures have been proven to be true. It took 28 years to obtain the complete solution of this problem, and the honor went to the German mathematicians O. Reutter for the conjecture and my friend Heiko Harborth from Braunschweig Technical University for the proof:

Reutter-Harborth's Theorem E13.2. (H. Harborth [H], 1974). In any set of n points in the plane the minimum distance r can occur at most $\lfloor 3n - \sqrt{12n - 3} \rfloor$ times.

Here, as usual, the symbol $\lfloor a \rfloor$ stands for the largest integer not exceeding a.

Harborth's result is best possible since his theorem's upper bound can be realized in an appropriate "extremal" configuration. In fact, 48 years after Erdős referred to the triangular lattice, in 1994 Yaakov S. Kupitz from Israel [Ku] showed that all extremal configurations can be obtained as sections of a triangular lattice. Figure 38.1 shows an example:

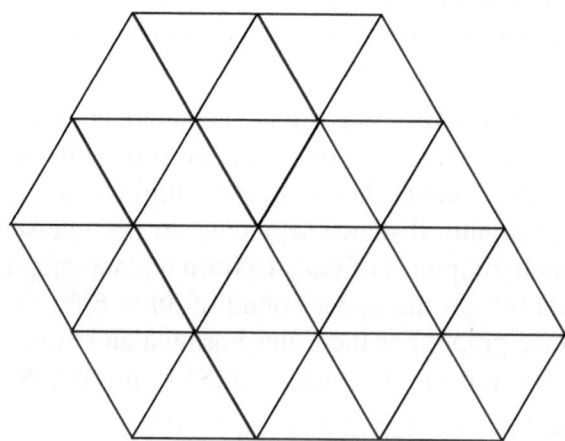

Fig. 38.1

Train B

Let us now switch to Train B and look at the multiplicity of the *diameter of a set*, i.e., the largest distance between the set's points. Problem 14.4(b), that I extracted from the same 1946 paper of Paul Erdős's [E0], has provided us with the upper bound n, where n is the number of points in the set. This upper bound is best possible. Paul simply writes, "It is easy to give n points where the maximum distance occurs exactly n times." Indeed, the configuration in Figure 38.2 provides a necessary example. In it, all drawn edges are of equal length (thus we have an equilateral triangle as a part of the configuration).

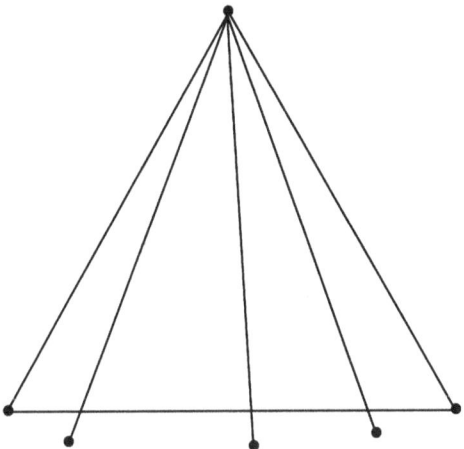

Fig. 38.2

Erdős [E0] suggests a further direction: "It would be interesting to have an analogous result for n points in k-dimensional space" – and immediately formulates the conjecture he received ("oral communication") from A. Vázsonyi:

Conjecture E13.2. (A. Vázsonyi, 1946 or before). In any n-element set in 3-dimensional space the maximum distance occurs at most $2n - 2$ times.

This conjecture was proven (and thus became a theorem) in 1956 independently by the great geometer and my friend Branko Grünbaum [Grü] and the Hungarian mathematician A. Heppes [He] and a year later by S. Staszewicz [St], although configurations in 3-space with $2n - 2$ diameters are still not well understood.

Paul Erdős then makes the following observation [E0]:

> If one could prove that in k-dimensional space the maximum distance [in an n-element set] cannot occur more than kn times, the following conjecture of Borsuk would be established: Each k-dimensional subset of diameter 1 can be decomposed into $k+1$ summands each having diameter < 1.

As you can see, for a long time Erdős believed in the positive resolution of this famous Borsuk Conjecture. In fact, in the foreword for [BS], [S14] Erdős wrote about trying to prove the Borsuk Conjecture circa 1940. This was not meant to be, for in July 1993 Jeff Kahn of Rutgers University and Gil Kalai of Hebrew University published [KK] a counterexample to the Borsuk Conjecture in dimension $k = 1326$. Read more about Borsuk Conjecture's aftermath in [S14].

If Erdős's statement above is correct, we can now conclude that

> There exists an n-element set in k-dimensional space such that the maximum distance occurs in it more than kn times.

There are more trains of great ideas in this classic 1946 paper of Paul Erdős. I will revisit the paper in the sequel of this book, since Problem 25.5 of the twenty-fifth anniversary Colorado Mathematical Olympiad in 2008 came from this wonderful paper as well!

E14. Squares in Square: The 1932 Erdős Conjecture[1]

Inspired by Problems 14.5(A), (B) and (C)

SECTION 1. OVERTURE

I will take you here on a journey through problems, conjectures, and results. My goal is to present a live fragment of mathematics and to show how mathematical train of thought travels. I hope you will enjoy the ride!

SECTION 2. ONCE UPON A TIME... or The 1932 Conjecture of Paul Erdős

Welcome aboard. Our train of thought is about to take off. We are in Budapest, Hungary, rolled back 68 years. In 1932, the 19-year old Paul Erdős posed the following problem. Inscribe in a unit square r squares that have no interior points in common. Denote by $f(r)$ the maximum of the sum of side lengths of the r squares. (We allow side lengths of squares to be zero.) The problem is to evaluate the function $f(r)$:

Open Problem E14.1. For every positive integer r find the value of $f(r)$.

In fact, the above formulation came about later, when Paul and I renewed efforts to settle the problem. Originally Paul formulated the following narrower but surprisingly difficult conjecture. When Paul

---

[1] A version of Explorations 14 and 15 appeared first as the author's talk at the International Congress on Mathematical Education in Tokyo, 2000, and published as [S10].

A. Soifer, *The Colorado Mathematical Olympiad and Further Explorations: From the Mountains of Colorado to the Peaks of Mathematics*, DOI 10.1007/978-0-387-75472-7_39, © Alexander Soifer, 2011   337

shared the conjecture with me in 1995, he offered a $50 prize for its first proof or disproof.

Fifty Dollar Squares in a Square Conjecture E14.2. (Paul Erdős, 1932). For any positive integer k,

$$f(k^2 + 1) = k.$$

The conjecture is still open today, in the year 2010, waiting, as Paul Erdős used to say, "for stronger arms, or, perhaps, brains" to be settled. But Paul and I made progress in a broader problem of describing the function $f(r)$.

SECTION 3. SIXTY-THREE YEARS LATER,
or Erdős-Soifer Results of 1995

First of all, Paul and I observed the following lower and upper bounds for $f(r)$. The symbol $\lfloor x \rfloor$ stands for the largest integer that is not greater than x.

Result E14.3. (P. Erdős and A. Soifer, 1995, [ES1]). The following inequality is true for any positive integer r:

$$\lfloor \sqrt{r} \rfloor \leq f(r) \leq \sqrt{r}.$$

Proof 1. *The Upper Bound.* The celebrated Cauchy Inequality (prove it on your own, it is fun to do :-) states that

$$\left(\sum_{i=1}^{r} a_i b_i \right)^2 \leq \left(\sum_{i=1}^{r} a_i^2 \right) \left(\sum_{i=1}^{r} b_i^2 \right).$$

Setting $b_i = 1$ for every $i = 1, 2, \ldots, r$, we get

$$\left(\sum_{i=1}^{r} a_i \right)^2 \leq \left(\sum_{i=1}^{r} a_i^2 \right) r. \tag{*}$$

Let r squares of side lengths a_i, $i = 1, 2, \ldots, r$, with no interior points in common be placed in a unit square. Then, the combined area of the r squares does not exceed the area of the unit square, i.e., $\sum_{i=1}^{r} a_i^2 \leq 1$, and we get from the inequality (*) above the required upper bound:

$$f(r) = \sum_{i=1}^{r} a_i \leq \sqrt{r}.$$

2. *The Lower Bound.* Surely, the function $f(r)$ is non-decreasing (do you see why?), therefore, $r \geq \lfloor \sqrt{r} \rfloor^2$ implies $f(r) \geq f(\lfloor \sqrt{r} \rfloor^2)$. Now let us partition the unit square into $\lfloor \sqrt{r} \rfloor^2$ congruent squares, each of the side length $\lfloor \frac{1}{\sqrt{r}} \rfloor$, and calculate the sum of side lengths of these $\lfloor \sqrt{r} \rfloor^2$ squares: we get $\lfloor \frac{1}{\sqrt{r}} \rfloor \times \lfloor \sqrt{r} \rfloor^2 = \lfloor \sqrt{r} \rfloor$. Observe that this partition and calculation demonstrate the inequality $f(\lfloor \sqrt{r} \rfloor^2) \geq \lfloor \sqrt{r} \rfloor$. By combining the two inequalities of this and preceding paragraphs, we get the required lower bound:

$$f(r) \geq f\left(\lfloor \sqrt{r} \rfloor^2\right) \geq \lfloor \sqrt{r} \rfloor$$

We are done! ∎

Result E14.3 has the following consequence:

Corollary E14.4. If $r = k^2$ for a positive integer k, then we get the equality

$$f(r) = k.$$

Corollary E14.4 allows us to see Erdős's Fifty Dollar Squares in a Square Conjecture in a slightly different light:

Fifty Dollar Squares in a Square Conjecture, Version 2, E14.5. (P. Erdős). At perfect square numbers $r = k^2$ (k is an integer) the function $f(r)$ does not increase:

$$f(k^2 + 1) = f(k^2).$$

Paul Erdős and I were able to prove that, amazingly, the function $f(r)$ is strictly increasing everywhere else! But to prove that we needed to find a much sharper lower bound for $f(r)$.

Result E14.6. (P. Erdős and A. Soifer, 1995, [ES1]). Any positive integer r can be presented in the form $r = k^2 + m$, where $0 \leq m \leq 2k$. Accordingly, the following inequalities hold:

(a) If $m = 2t + 1$, where $0 \le t < k$, then $f(r) \ge k + \frac{t}{k}$;
(b) If $m = 2t$, where $0 \le t \le k$, then $f(r) \ge k + \frac{t}{k+1}$.

Proof Given a positive integer r, we can present it in a form

$$r = k^2 + m \text{ with } 0 \le m \le 2k.$$

Indeed, it suffices to choose $k = \lfloor \sqrt{r} \rfloor$. If r is a perfect square $r = k^2$, then $m = t = 0$, and Corollary E14.4 provides the exact value $f(r) = k$, which is a part of the required inequality (b). We can assume now that r is *not* a perfect square, i.e., $m \ne 0$. The parity of m dictates two cases.

(a) $m = 2t + 1$ and $0 \le t < k$. Let us first partition the unit square into k^2 congruent squares (we get a $k \times k$ square grid, call it G), and then replace a $t \times t$ subgrid of the grid G with a $(t + 1) \times (t + 1)$ square grid of the same total size as the removed subgrid (see Figure 39.1).

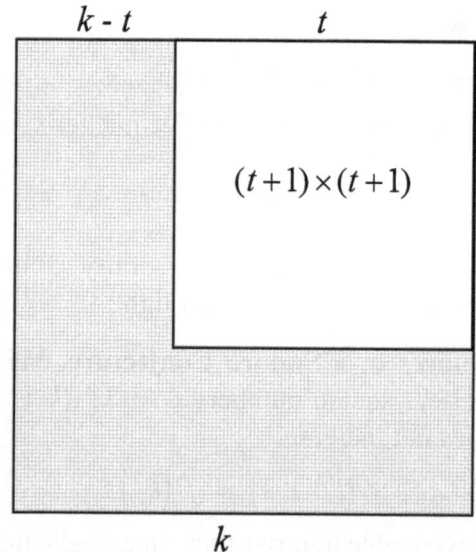

Fig. 39.1

We end up with a partition of the unit square into

$$k^2 - t^2 + (t + 1)^2 = k^2 + 2t + 1$$

little squares, some of which [the original ones] have side length $\frac{1}{k}$, and others [the squares of the inserted $(t + 1) \times (t + 1)$ square grid] side length $\frac{t}{k(t+1)}$. Let us calculate the sum of side lengths of all these $k^2 + 2t + 1$ little squares. We get

$$\frac{1}{k}k^2 - \frac{1}{k}t^2 + \frac{t}{k\,(t + 1)}\,(t + 1)^2 = k + \frac{t}{k}.$$

This partition and calculation deliver the following lower bound for $f(r)$:

$$f\,(r) \geq k + \frac{t}{k}.$$

(b) $m = 2t$ and $0 < t \leq k$. We first partition the unit square into $(k + 1)^2$ congruent squares (we get a $(k + 1) \times (k + 1)$ square grid, call it G), and then replace a $(k - t+1) \times (k - t+1)$ subgrid of the grid G with a $(k - t) \times (k - t)$ square grid of the same total size as the removed subgrid (see Figure 39.2).

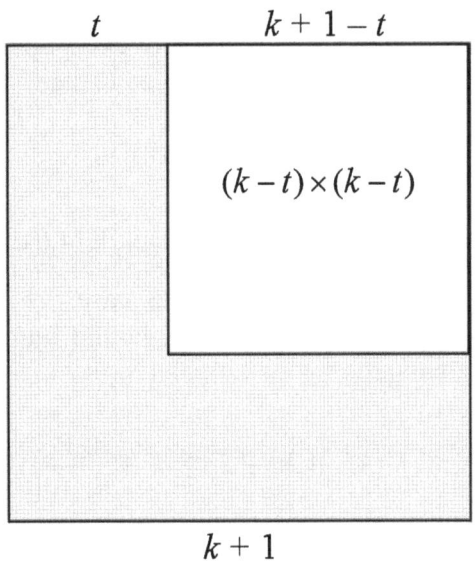

Fig. 39.2

We end up with a partition of the unit square into

$$(k + 1)^2 - (k - t + 1)^2 + (k - t)^2 = k^2 + 2t$$

little squares, some of which [the original ones] have side length $\frac{1}{k+1}$, and others [the squares of the inserted $(k-t) \times (k-t)$ square grid] side length $\frac{k-t+1}{(k+1)(k-t)}$. Let us calculate the sum of side lengths of all these $k^2 + 2t$ little squares, we get

$$
\frac{1}{k+1}(k+1)^2 - \frac{1}{(k+1)(k-t+1)}(k-t+1)^2
$$
$$
+ \frac{k-t+1}{(k+1)(k-t)}(k-t)^2 = k + \frac{t}{k+1}.
$$

This partition and calculation deliver the following lower bound for $f(r)$:

$$
f(r) \geq k + \frac{t}{k+1}.
$$

Done! ■

Result E14.7.[2] (P. Erdős and A. Soifer, 1995, [ES1]). The function $f(r)$ is strictly increasing everywhere except possibly at perfect square points, i.e., if $r \neq k^2$ for any integer k, then

$$
f(r+1) > f(r).
$$

Proof Once again the parity of m and Result E14.6 dictate two cases.

(a) $m = 2t + 1$ and $0 \leq t < k$. In this case $t + 1 \leq k$, and by substituting $t + 1$ for t in the lower bound found in part (b) of Result E14.6, we get:

$$
f(k^2 + 2t + 2) \geq k + \frac{t+1}{k+1}.
$$

[2] This result was also obtained independently by D.G. Rogers and S. Suen (fax to P. Erdős of March 9,1995).

This inequality and Result E14.3 deliver the necessary chain of inequalities:

$$f(r) = f\left(k^2 + 2t + 1\right) \leq \sqrt{k^2 + 2t + 1} < k + \frac{t+1}{k+1}$$

$$\leq f\left(k^2 + 2t + 2\right) = f(r+1).$$

(b) $m = 2t$ and $0 < t \leq k$. By using Result E14.3 and the lower bound found in part (a) of Result E14.6 above, we get the necessary chain of inequalities:

$$f(r) = f\left(k^2 + 2t\right) \leq \sqrt{k^2 + 2t} < k + \frac{t}{k} \leq f\left(k^2 + 2t + 1\right)$$

$$= f(r+1).$$

Result E14.7 is proven. ∎

Note: In the proof above I omitted a demonstration of two inequalities: $\sqrt{k^2 + 2t + 1} < k + \frac{t+1}{k+1}$ and $\sqrt{k^2 + 2t} < k + \frac{t}{k}$. I hope their verification will be a welcome exercise in secondary school algebra for you :-).

Paul Erdős and I believed that the lower bounds in Result E14.6 were quite good, and conjectured that they just may be the best possible:

Conjecture E14.8. (P. Erdős and A. Soifer, 1995, [ES1]). Any positive integer r can be represented in the form $r = k^2 + m$, where $0 \leq m \leq 2k$. Accordingly, we conjecture the following equalities:

(a) If $m = 2t + 1$, where $0 \leq t < k$, then $f(r) = k + \frac{t}{k}$;
(b) If $m = 2t$, where $0 \leq t \leq k$, then $f(r) = k + \frac{t}{k+1}$.

We also observed that our examples in Result E14.6 completely tile the unit square, and thus posed the following open problem:

Open Problem E14.9. (P. Erdős and A. Soifer, 1995, [ES1]). Is it true that for any positive integer r, the value of $f(r)$ can be attained by a set of r squares that form a complete tiling of the unit square by themselves or with an addition of at most *one* extra square?

For years Paul Erdős (1913-1996) and I worked on the book of his incredible open problems. I hope to finish this joint book, *Problems of pgom Erdős* by 2012 or so. It will be published by Springer.

E15. From Squares in a Square to Clones in Convex Figures

**Inspired by Problems 14.5(A), (B), and (C)
and by the Erdös-1932 Conjecture you met in E14**

SECTION E15.1. A NEW SQUARES IN SQUARE PROBLEM

As I thought about Paul Erdős's problem, it appeared natural for me to pose a "dual" problem, and thus give birth to the *New Squares in a Square Problem*.

Let $\square$ stand for a square and $r > 1$ a positive integer. Denote by $S(\square, r)$ the smallest area of a square Q such that any r squares whose areas add up to at most 1, can be packed in Q (i.e., embedded in Q with no interior points in common).

You proved with me in the Olympiad Problems 14.5(A), (B), and (C) that for any r in the range $2 \leq r \leq 5$, $S(\square, r) = 2$, and *I* conjectured that in fact, $S(\square, r) = 2$ for *any* positive $r > 1$:

Squares in Square Conjecture E15.1. (A. Soifer, 1995, [S7]). For any positive integer $r > 1$,

$$S(\square, r) = 2.$$

Two years have passed since I created this conjecture. In May 1997, I was in Lincoln, Nebraska grading papers of the USA Mathematical Olympiad, together with other members of the USA Mathematics Olympiad Subcommittee. During a break, I put the New Squares in Square Conjecture on the board. Later the same day Richard Stong told me "I proved your conjecture." Indeed, he did!

A. Soifer, *The Colorado Mathematical Olympiad and Further Explorations: From the Mountains of Colorado to the Peaks of Mathematics*, DOI 10.1007/978-0-387-75472-7_40, © Alexander Soifer, 2011

Richard devised a simple "greedy" algorithm and a nice, clever proof that his algorithm works[1], and thus New Squares in Square Conjecture became an important result, which I happily published in *Geombinatorics*:

New Squares in Square Result E15.2. (R. Stong, 1997, [Sto]). Any finite set of squares of the combined area 1 can be packed in a square of area 2.

In 1984 Richard Stong won the William Lowell Putnam Competition. He is a professor of mathematics at Rice University.

Later I discovered that Problem E15.1 and Result E15.2 were not new, and although Stong's proof was better, he was preceded by 30 years by J.W. Moon and Leo Moser of Edmonton, Alberta, Canada [MoM]. Ecclesiastes (1:9-14 NIV) comes to mind:

> What has been will be again, what has been done will be done again; there is nothing new under the sun.

The good thing did take place: I brought a new excitement and new players to the problem. Moreover, I was already riding the train of further thoughts, the one, it seems, no one has traveled before.

SECTION E15.2 PACKING DISCS IN DISC

I conjectured [S8] that the identical result was true for circular discs (I will use here the word "disc" to mean circular disc). This 1998 conjecture is still open today, 12 years later:

Discs in Disc Conjecture E15.3. (A. Soifer, [S8]). Any finite set of discs of combined area 1 can be packed in a disc of area 2.

In writing these lines I asked myself, can we prove any upper bound – even though the bound 2 is out of reach at the moment?

I can prove an easy one for you:

Upper Bound for Disc Packing E15.4. Any finite set of discs of combined area 1 can be packed in a disc of area 4.

---

[1] Space considerations prevent me from including this proof here, but I recommend everyone to read it in *Geombinatorics*.

Proof Given a finite set of discs of combined area 1. Circumscribe a square about each of the discs. The combined area of squares is $4/\pi$. By result E15.2 we can pack these squares in a square Q of area $8/\pi$. Finally Q can be inscribed in a disc of area $8/\pi \times \pi/2 = 4$. ∎

In working with similar triangles we encounter issues that do not exist for squares and circular discs – limitations on the way clones are embedded. We can limit embeddings to translations, and thus define a function $S_T(\Delta)$. Or we can place no limitations on embeddings at all and end up with our original $S(\Delta)$. Of course, $S(\Delta) \leq S_T(\Delta)$. It is not at all obvious whether these two values are equal!

In 1995 T. J. Richardson calculated the easier of the two values:

Packing Triangles in a Triangle E15.5. (T.J. Richardson, 1995 [Ri]). Any finite set of similar to each other triangles of combined area 1 can be packed in a triangle similar to them of area 2.

On January 27, 2009, the Polish geometer Janusz Januszewski informed me that he had calculated the harder value $S_T(\Delta)$ [J2]:

Embedding Triangles in a Triangle by Translations Alone E15.6. (J. Januszewski, 2009, [J2]). For any triangle Δ, $S_T(\Delta) = 2$, i.e., any finite set of similar triangles of combined area 1 can be packed in a triangle similar to them of area 2 by translations alone.

SECTION E15.3 PACKING CLONES IN CONVEX FIGURES

Let us roll the time back 11 years. By 1998 I felt it was time to generalize these observations to include *all* geometric figures in our "games." I got busy.

Given figures f and F; it is convenient to call a figure f an *F-clone* if f is similar to F.

Given a figure F, let $S = S(F)$ be the minimum real number such that any finite set of F-clones of combined area 1 can be packed in an F-clone of area S.

Results E15.2 and E15.5 can be written in this notation as follows:

$$\text{For a square } \square, S(\square) = 2;$$
$$\text{For any triangle } \Delta, S(\Delta) = 2.$$

However, it is easy to see that the numbers $S(F)$ are not even bounded if we impose no limitation on figures F in the study:

Result E15.7. (A. Soifer, 1998, [S9]). For any number r, there is a figure F such that $S(F) > r$.

Proof Indeed, for any r, we can construct a cross C thin enough that only one of the two C-clones of area $\frac{1}{2}$ can be inscribed in a C-clone of area r (Figure 40.1). ■

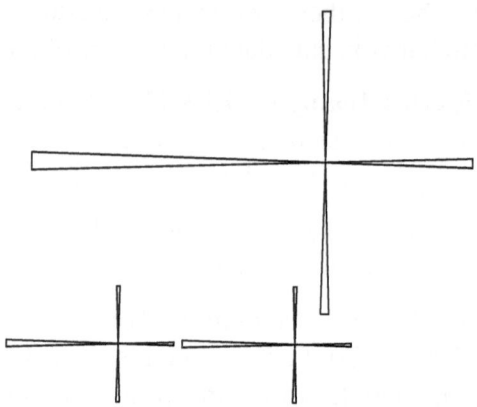

Fig. 40.1

Thus, it makes sense to limit the scope of our games to *convex* figures. The main problem then can be formulated as follows:

Main Open Problem E15.8. (A. Soifer, 1998, [S9]). For any convex figure F, find $S(F)$.

This is a very difficult problem that in full generality may withstand centuries – "but we will see," as Paul Erdős used to jokingly remark. However, partial solutions are possible and most welcome!

SECTION E15.4 THE MAIN CONJECTURE

I hope you have enjoyed the ride. Out of the window of our train of mathematical thought you may have noticed the terrain that has been continuously changing, with one problem giving birth to another.

The journey is not over: in fact, it has only begun. We mastered the packing field somewhat, and are now ready to commence a search for its essence, a result encompassing all convex figures. In 1998 I came up with a brave conjecture:

Clones in Convex Figures Conjecture E15.9. (A. Soifer, 1998). For *any* convex figure F, any set of F-clones $F_1, F_2, \ldots, F_n$ whose areas add up to at most 1 can be packed in a clone F_0 of area 2, i.e.,

$$S(F) \leq 2.$$

However, when I wrote this conjecture up for *Geombinatorics* [S9], I inadvertently put it as

$$S(F) = 2.$$

As the Russian proverb has it, *There is no bad without some good in it!* Three years later, in 2001, the Slovak geometer Pavel Novotný constructed a counterexample to the published equality $S(F) = 2$:

Novotný's Example E15.10. ([N], 2001). For a rectangle R_0 of size $\sqrt[8]{\frac{3}{2}} \times \sqrt[8]{\frac{2}{3}}$ we get

$$S(R_0) = \sqrt{\frac{8}{3}} < 2.$$

Novotný understood my typo, for he wrote, "Soifer's conjecture could be changed to $S(F) \leq 2$."

A year later, in 2002, Janusz Januszewski [J1] beautifully completed the above result of Novotný:

Januszewski's Theorem E15.11. ([J1], 2002). For any rectangle R, $S(R) \leq 2$. Moreover, $S(R) = 2$ if and only if the rectangle R is a square.

Janusz gave the main conjecture its final proper attribution:

The Soifer-Novotný Conjecture E15.12. ([J1], 2002). For any convex figure F, $S(F) \leq 2$.

Finally, Janusz Januszewski posed a natural problem that is a particular case of my Problem E15.8:

Januszewski's Problem E15.13. [J1]. Classify convex figures F for which $S(F) = 2$.

He, among others, noticed that, perhaps, the Soifer-Novotný Conjecture can be generalized to n-dimensional Euclidean spaces:

Januszewski's Conjecture E15.14. [J1]. Let F be a convex body in an n-dimensional Euclidean space. Then any set of F-clones can be packed in an F-clone of volume 2^{n-1}.

I would like to know the minimum value of $S(F)$:

Open Problem E15.15. Find min $S(F)$ over all convex figures F and classify figures for which this minimum is attained.

Most of these series of results appeared on pages of *Geombinatorics*, a quarterly dedicated to problem posing essays in combinatorial and discrete geometry (hence its title). They were met with interest by the mathematical community. Peter Winkler of Dartmouth University included a discussion of Conjectures E15.3 and E15.12 and Result E15.2 in his book [W, pp. 146 and 157]:

> This lovely conjecture is due to Alexander Soifer of the University of Colorado, Colorado Springs. It and its relatives have been the subject of a dozen of articles in the journal *Geombinatorics*; it is known, for example, that squares of total area 1 can be packed into a square of total area 2. The generalization to higher dimension was suggested by your author, among others; the case of two balls, each of volume 1/2, shows that 2^{d-1} is best possible.

I hope you have enjoyed your ride on this train of mathematical thought!

E16. Olde Victorian Map Colouring

Inspired by problem 15.5

Let us start with the history of the arguably second most famous problem in the entire history of mathematics. It commences in Victorian London in the year 1852, when the 20-year old Francis Guthrie created The Four-Color Conjecture (4CC), and continues for 124 years, when in 1976 Kenneth Appel and Wolfgang Haken, with the assistance of John Koch and over 1200 hour of mainframe computer time converted 4CC into 4CT – The Four Color Theorem. A second proof had to wait another 20 years: in 1997 Neil Robertson, Daniel Sanders, Paul Seymour and Robin Thomas also proved 4CT. Both proofs required an essential use of computing.

Why then did we need the second proof?

It was better than the first!

What does "better" mean here?

The new proof featured a clean separation of computer and human contributions; moreover, it was fully verifiable and mathematically much simpler. Read more about the exciting history of this problem in *The Mathematical Coloring Book* [S11].

Here I would like you to look at a couple of problems, stemming from Alfred Bray Kempe's 1879 [K] and Peter Guthrie Tate's 1880 [T] papers.

I am sure you realize that in map coloring, the shape of a circle is of no consequence. We can replace circles in Problem 15.5(A) by their

A. Soifer, *The Colorado Mathematical Olympiad and Further Explorations:*
From the Mountains of Colorado to the Peaks of Mathematics,
DOI 10.1007/978-0-387-75472-7_41, © Alexander Soifer, 2011

continuous one-to-one images, called *simple closed curves*, because the Jordan Curve Theorem holds for them all:[1]

Jordan Curve Theorem E16.1. A simple closed curve in the plane divides the plane into two regions (inside and outside).

E16.2 Prove that a map formed in the plane by finitely many simple closed curves is 2-colorable.

We can replace simple closed curves by straight lines, or a combination of the two:

E16.3 Prove that a map formed in the plane by finitely many straight lines is 2-colorable (Figure 41.1).

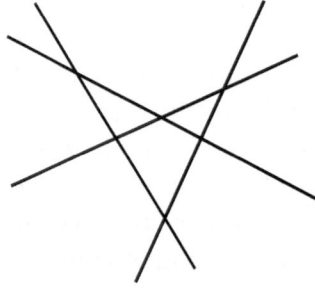

Fig. 41.1

An inductive proof is well known, but, as is usually the case with proofs by induction, does not provide any insight. I found a "one-line" proof that takes advantage of similarity between simple closed curves and straight lines.

Proof Attach to each line a vector perpendicular to it (Figure 41.2). Call the half-plane *inside* if contains the vector, and *outside* otherwise. Repeat the solution of Problem 15.5(A) word-by-word to complete the proof (Figure 41.3). ∎

E16.4 Prove that a map formed in the plane by finitely many simple closed curves and straight lines is 2-colorable.

---

[1] see its proof, for example, in [BS], [S14].

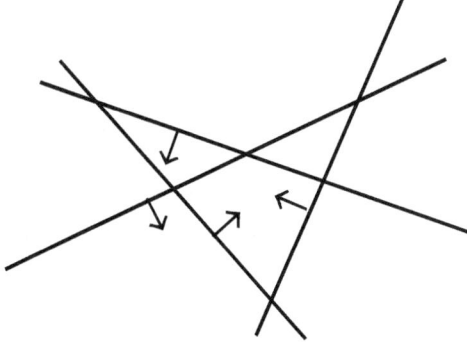

Fig. 41.2

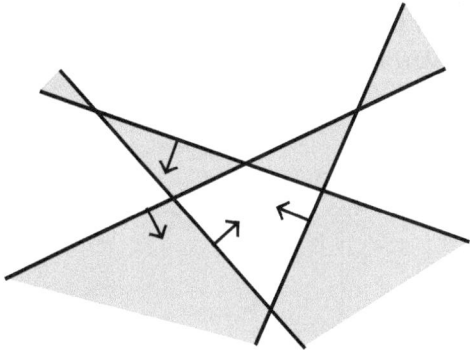

Fig. 41.3

So, what is common between simple closed curves and straight lines? What allows a 2-coloring to exist? Of course, it is the fact that each of them partitions the plane into two regions. But an alternative observation is possible. Each vertex in the maps above is a result of the intersection of two or more curves or lines, and therefore has even degree (the *degree of a vertex* is the number of lines emanating from it)! This fact first appears in print on the last page of the 1879 paper by Alfred Bray Kempe (1849-1922) in which he attempts to prove 4CC [K].

Kempe's Two-Color Theorem E16.5. (A. B. Kempe, 1879, [K]). A map is 2-colorable if and only if all its vertices have even degree.

Let us take another look at the map M formed by circles in Figure 15.7. We can construct the *dual graph*[2] $G(M)$ of the map M as follows: we represent every country by a vertex (think of the capital city), and call two vertices adjacent if the corresponding two countries are adjacent, i.e., share a common boundary (not just a point or finitely many points).[3] The dual graph $G(M)$ of the map M in Figure 15.7 is shown in Figure 41.4 (I bent and stretched edges to make the graph look aesthetically pleasing).

Observe: the dual graph $G(M)$ of any map M is *planar*, i.e., we can draw its edges through common boundaries of adjacent countries so that the edges will have no points in common except the vertices of the graph.

Now the problem of coloring maps can be translated into the language of coloring vertices of planar graphs.

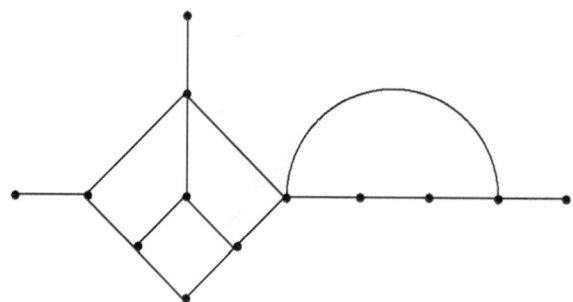

Fig. 41.4

Kempe's Two-Color Theorem E16.6. (*In Graph-Theoretical Language*). The chromatic number $\chi(G)$ of a graph G does not exceed 2 if and only if G contains no odd cycles.

It is now time to look at Problem 15.5(B), which, as you recall, is due to Peter Guthrie Tait. In one year, 1880, Tait published a paper, withdrew it, and replaced it with a one-page "abstract," which he

---

[2] See the definition of a graph at the beginning of the solution of Problem 7.6.

[3] The idea of the dual graph of a map was one of the first ideas of graph theory: Leonard Euler used it in 1736 to solve the Problem of Bridges of Königsberg. The language of maps was universally used by the first researchers of the 4CC. Yet, it is interesting to notice that while Kempe used the language of maps in the main body of his 1879 paper [Kem2], he did describe the construction of the dual graph on the last page of this paper.

expanded to an article [T]. This paper contains some amusing statements, for example [T, p. 657]:

> The difficulty in obtaining a simple proof of this theorem originates in the fact that it is not true without limitation.

One can paraphrase this to say, "It is difficult to prove what is not true." Indeed, very much so! The Tait paper, however, also contains brilliant observations, such as what we call Tait's Equivalence (the statement of Problem 15.5(B) plus its converse).

Let p be a point on a boundary in a map M. The *degree* of a point p in a map is the number of boundary curves that emanate from p. We will call a point *essential* if its degree is at least 3. We are ready for the next problem.

Problem E16.7. (P. H. Tait). Prove the converse of the result of Problem 15.5(B): Let M be a 4-colorable map, with all essential points having degree 3. Prove that the set of essential points together with all boundary lines of M forms a triple-delight (see the definition in Problem 15.5(B)).

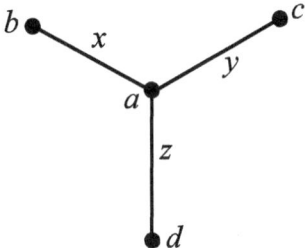

Fig. 41.5

Proof [T]. Let the vertices of a planar 3-regular graph G be 4-colored in colors a, b, c and d. We then color the edges in colors x, y, and z as follows: an edge is colored x if it connects vertices colored a and b, or c and d; an edge is colored y if it connects vertices colored a and c, or b and d; and an edge is colored z if it connects vertices colored a and d, or b and c. You can easily verify that a proper edge coloring is thus obtained, i.e., no adjacent edges are assigned the same colors. In view of symmetry, it suffices to show this for the edges incident to a vertex colored a, which is demonstrated in Figure 41.5. ∎

I will touch on map coloring again in *Exploration* E20 – see you there!

E17. Achievement Games: Is Snaky a Loser?

Inspired by Problems 17.3(A) and 17.3(B)

Problems 17.3(A) and 17.3(B) come from "achievement" games invented apparently in 1976 by the famous graph theorist Frank Harary, author of over 700 papers and the well known (I enjoyed reading it in Russian translation) 1969 book *Graph Theory*. Frank wrote an article [Har] about these games for *Geombinatorics* that was "Dedicated to Martin Gardner, who gave Snaky its name." Therefore I have a rare opportunity and distinct pleasure of giving the microphone to the inventor. Frank Harary explains [Har]:

> Generalizations of tic tac toe, the best known board game in the world, have been proposed in many directions. My favorite among these involves the family of plane configurations which have been called square-cell animals or polyominoes. Given an animal L and the infinite mesh M as the playing board, two players A and B alternately color green and red respectively the cells of M. Thus, A first colors a cell of M green, then B colors another cell red, etc. In the L-achievement game, the first player who has formed in his color a copy of L is the winner. As the outcome with rational play is either a victory for A or a draw, we call generic animal L a winner if A can form a green L in M regardless of the moves made by B.

In this terminology, Problems 17.3(A) and (B) prove that the 5-cell cross and the 2×2 square are losers.

Problem E17.1. Prove that the animal in Figure 42.1 is a loser.

A. Soifer, *The Colorado Mathematical Olympiad and Further Explorations: From the Mountains of Colorado to the Peaks of Mathematics*, DOI 10.1007/978-0-387-75472-7_42, © Alexander Soifer, 2011

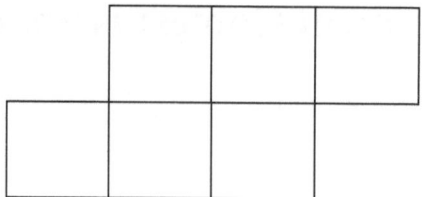

Fig. 42.1

Proof. Tile the plane by dominoes as shown in Figure 42.2.

Fig. 42.2

Observe that no matter how our animal is placed on the square grid, it will cover a domino completely. Hence the strategy for player *B* is clear: as soon as player *A* colors a cell green, *B* has to color red the second cell of the same domino. ∎

Try to prove another animal loser on your own:

Problem E17.2. Prove that the animal in Figure 42.3 (a linear pentomino) is a loser.

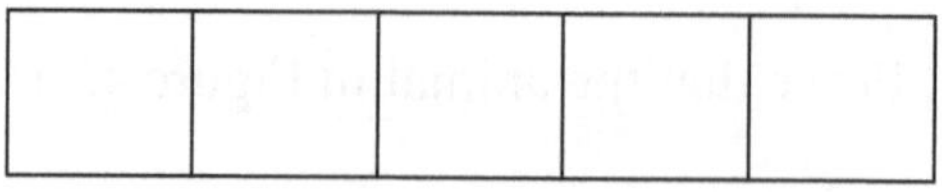

Fig. 42.3

Frank Harary continues [Har]:

> There exists exactly one animal, called Snaky [Figure 42.4], whose winning status is not yet known. A prize of U.S. $100 will be awarded to the first person who definitely establishes on or before 31 December 1995 whether or not Snaky is a winner.

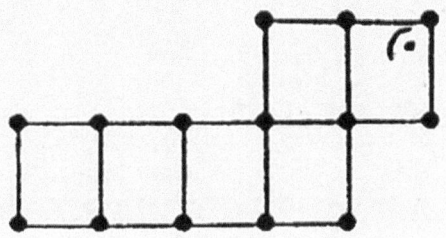

Fig. 42.4 "The Snaky animal," a portrait by Frank Harary [Har]

The deadline has obviously passed. However, Snaky is still awaiting a snake handler! I would conjecture, Snaky is a loser.

The Snaky Open Problem E17.3. (F. Harary). Is Snaky a loser?

E18. Finite Sets of Infinite Fun: Three Classic Problems

Inspired by Problem 19.4

Did you enjoy the Olympiad Problem 19.4? Then brace yourself for much more fun! While all explorations in different ways are dear to my heart, this small collection may provide the greatest enjoyment of all. Some of these problems have been solved, others are still awaiting their conqueror, but all of them satisfy the definition of "classic" problems of mathematics.

Classic I

The first of the theorems to be discussed in this Exploration was discovered in 1913 by the Austrian mathematician Eduard Helly (Vienna, 1884 – Chicago, 1943). I suggest you to try to prove on your own the Helly Theorem for the plane in two stages:

Problem E19.1. Four *convex* figures are given in the plane. Prove that if every three of them have non-empty intersection, then the intersection of all four figures is non-empty as well.

The Helly Theorem for the Plane E19.2. Given a finite family $F_1, \ldots, F_s$ of convex figures in the plane, prove that if every three of them have non-empty intersection, then the intersection $F_1 \cap \ldots \cap F_s$ of all of these figures is non-empty as well.

You will find an exciting story of the "proliferation" of the Helly Theorem in some applications of the Helly Theorem in [BS] and

A. Soifer, *The Colorado Mathematical Olympiad and Further Explorations: From the Mountains of Colorado to the Peaks of Mathematics*, DOI 10.1007/978-0-387-75472-7_43, © Alexander Soifer, 2011

[S14]. See also some nice results stemming from the "marriage" of the Helly Theorem and the Ramsey Principle in *The Mathematical Coloring Book* [S11].

However, if you really-really enjoyed the Helly Theorem, you ought to find and read the classic paper *Helly's Theorem and Its Relatives* [DGK] written by three fabulous mathematicians: Ludwig Danzer, Branko Grünbaum, and Victor Klee. In my young years, I read it in the Russian translation.

Classic II

Our second classic theorem was discovered by two young Hungarian mathematicians little known at the time, who had a great future ahead of them.

A sequence $a_1, a_2, ..., a_k$ of real numbers is called *monotone* if it is increasing, i.e., $a_1 \leq a_2 \leq ... \leq a_k$, or decreasing, i.e., $a_1 \geq a_2 \geq ... \geq a_k$ (we use the weak versions of these definitions that allow equalities of consecutive terms).

The Erdős-Szekeres Monotone Subsequence Theorem E19.3. [ESz]. Any sequence $S : a_1, a_2, ..., a_r$ of $r > mn$ real numbers contains a decreasing subsequence of more than m terms or an increasing subsequence of more than n terms.

A quarter century later, in 1959, A. Seidenberg of the University of California, Berkeley, found a brilliant short proof of theorem E19.3, thus giving it a true Olympiad-like appeal.

Proof by A. Seidenberg [Sei]. Assume that the sequence $S: a_1, a_2, ...,$ a_r of $r > mn$ real numbers has no decreasing subsequence of more than m terms. To each a_i assign a pair of numbers (m_i, n_i), where m_i is the largest number of terms of a decreasing subsequence beginning with a_i and n_i the largest number of terms of an increasing subsequence beginning with a_i. This correspondence is an injection, i.e., distinct pairs correspond to distinct terms $a_i, a_j, i < j$. Indeed, if $a_i \leq a_j$ then $n_i \geq n_j + 1$, and if $a_i > a_j$ then $m_i \geq m_j + 1$.

We get $r > mn$ distinct pairs (m_i, n_i) (they are our pigeons) and m possible values (they are our pigeonholes) for m_i, since $1 \leq m_i \leq m$. By the Pigeonhole Principle, there are at least $n+1$ pairs (m_0, n_i) with the same first coordinate m_0. Terms a_i corresponding to these pairs (m_0, n_i) form an increasing subsequence! ∎

Erdős and Szekeres note that the result of their Theorem E19.3 is best possible – prove it on your own:

Problem E19.4. ([ESz]) Construct a sequence of mn real numbers such that it has no decreasing subsequence of more than m terms *and* no increasing subsequence of more than n terms.

Classic III

The third classic problem I am offering to you here is not a theorem, at least not yet. It is 38 years old and yet nobody knows how to prove it. Be the first!

Once upon a time – as I heard in September 1972 – at a party in Boulder, a city in my state of Colorado, the three mathematicians Paul Erdős, Vance Faber, and László Lovász formulated what they thought was an easy problem:

The Erdős-Faber-Lovász Conjecture E19.5. The union of n pairwise edge-disjoint complete graphs on n vertices is n-colorable.

You can appreciate the precision of the conjecture by proving the following result the authors observed right away:

Counterexample E19.6. [E3]. The conjecture fails if $n+1$ graphs are given instead of n.

In 1973 Paul Erdős writes:

> It is surprising that this simple conjecture seems to be rather difficult.

Soon Paul offered "100 dollars for a proof or disproof" [E4], and the prize for the first solution rose to $500 in the early 1980s [E5]. In fact, Erdős opens his paper *Some recent problems and results in graph theory* [E5] with the Erdős-Faber-Lovász Conjecture, as it slowly became one of his favorite open problems!

E19. From King Arthur to Mysteries of Hamiltonian Graphs

Inspired by problem 19.5

As you will recall, in the solution of Problem 19.5 we introduced for our convenience the notion of *Hamiltonian cycle*, or a walk along the edges through all vertices of a graph, that does not repeat a vertex (and therefore, does not repeat any edge) until it ends at the vertex it started from. When such a Hamiltonian cycle exists, the graph is called *Hamiltonian*.

Graph theory also knows the dual notion of an *Eulerian cycle*: a walk along all edges of a graph without repeating an edge until it ends at the vertex it started. When such an Eulerian cycle exists, the graph is called *Eulerian* (sometimes the requirement of ending at the starting point is relaxed).

Two similar notions bear two similar problems:

*) find a criterion for a graph to be Eulerian;
**) find a criterion for a graph to be Hamiltonian.

Amazingly, one of these problems is very easy, and the other is unsolved! As a result, one of these problems is an exercise, while solving another would likely land your name on the front page of the *New York Times*!

Can you guess which one is easy? It is not immediately obvious, is it? All right, I will disclose it for you. Enjoy proving on your own the following criterion, found by the great Leonard Euler in 1736, and by doing so laying the foundation for Graph Theory.

A graph is called *connected* if there is a walk between any two vertices of the graph along a series of its edges.

A. Soifer, *The Colorado Mathematical Olympiad and Further Explorations: From the Mountains of Colorado to the Peaks of Mathematics*, DOI 10.1007/978-0-387-75472-7_44, © Alexander Soifer, 2011

Euler's Criterion E19.1. (L. Euler [Eu], 1736). A connected graph is Eulerian if and only if the degree of each its vertex is even.

That is all; the Eulerian graph criterion is this simple! The problem with Hamiltonian graphs is a totally different matter. We do not have a criterion. Furthermore, we do not even have a conjecture – mathematicians have no idea what to expect from such a criterion in spite of a major effort by many researchers. We do have sufficient conditions, and necessary conditions – however, they clearly do not meet to form a criterion!

Our *King Arthur* Problem 19.5 is a particular case of the following sufficient condition, found by one of the creators of modern Graph Theory, Norwegian born and educated Yale University Professor Øystein Ore (1899-1968).

Ore's Sufficient Condition E19.2. (Ø. Ore [O], 1960). Given a graph G on $n > 2$ vertices. If any two non-adjacent vertices u, v of G satisfy the inequality $\deg u + \deg v \geq n$, then G is Hamiltonian.

I leave the proof to you – especially since we have discovered all the ideas that work here at King Arthur's Court, while solving Problem 19.5.

So, the ultimate open problem is this:

Open Problem E19.3. Find a criterion for a graph to be Hamiltonian.

Try to solve this problem, think about it from time to time – any advancement would be of great interest – but do not bet your entire life on such a mysterious problem!

E20. Exploring Map Coloring

Inspired by problem 20.5

The point of this problem is to "contradict" the Four-Color Theorem. In fact, problem 20.5 illuminates this famous theorem in a novel way.

E20.1 The Four-Color Theorem (Appel-Haken-Koch, 1976). Four colors suffice to color any plane map.

I expected that educated students would know the Four-Color Theorem, and thus misjudge Problem 20.5 and try to prove a negative answer to it. This would have taken a very long time, since proving what isn't true usually does :-).

In the terminology of Problem 20.5, the Four-Color Theorem would read as follows:

Once the explorer draws a complete map, the mapmaker can always color it with four colors.

Aleksandr Kovaldzhi's brilliant Problem 20.5 illustrates that there is no commutativity of the operations of drawing countries and coloring them. If the explorer creates one country at a time, and the mapmaker must color it, no number of colors suffices if the explorer is clever enough and the number of countries is large enough!

Let us now go back to the Four-Color Theorem (4CT). We know (4CT) that every finite map on the plane is 4-colorable. What about maps with infinitely many countries? This sounds like a natural question, which I have heard from various people at various times. In particular, Peter Winkler, then Director of Fundamental Mathematics Research at Bell Labs and now professor of mathematics at Dartmouth,

A. Soifer, *The Colorado Mathematical Olympiad and Further Explorations: From the Mountains of Colorado to the Peaks of Mathematics*, DOI 10.1007/978-0-387-75472-7_45, © Alexander Soifer, 2011

asked me this question on October 11, 2003 right after my talk at Princeton-Math. One proof, limited to an infinite sequence of countries, was given to me the following year.

In late December of 2004 I was invited to give two talks at the Mathematical Sciences Research Institute in Berkeley, California. There my old friend Professor Gregory Galperin showed me his proof of 4CT for infinitely many countries. His proof worked only for maps with countable number of countries. It was strikingly beautiful.

Countable Map Coloring Theorem E20.2. Every plane map with countably many countries $C_1, C_2, \ldots, C_n, \ldots$ is 4-colorable.

It is late (in this book's time :-)), and I do not wish to deny you the pleasure of trying to discover a proof on your own. You will find Grisha Galperin's proof in Chapter 23 of *The Mathematical Coloring Book* [S11].

The theorem is true for uncountably many countries as well – read its proof in [S11], in which you will find much more fun of the coloring kind – and the lives of those who invented it!

Infinite Map Coloring Theorem E20.3. Every plane map with infinitely many countries is 4-colorable.

Part V
Winners Speak: Reminiscences in Eight Parts

CMO is the best kind of math competition, because it tests creativity and intelligence rather than mere knowledge and training.

– James Carroll

Winner of the Literary Award 1999 & 2001, and Third Prize 2000

Introduction to Part V

While preparing this new much-expanded edition, a thought visited me: Let me offer a microphone (or is it a writing feather or a computer keyboard?) to the past winners, to share what they thought about the Colorado Mathematical Olympiad, how it affected their lives, and what their lives AD (*After Departure* from high school and the Olympiad) have been like.

I will start with a newspaper article that interviews some of the winners, then will reproduce the 1989 letter of our first ever winner of the Literary Award, followed by letters of two of the Olympiad's silver medalists, and conclude with recent reminiscences of four gold medal winners. Hence, *Reminiscences in Eight Parts!*

I have just reread the winner's letters. The depth and breadth of their experiences, thoughts, observations, and careers is breathtaking! There is no higher calling than to help these young talents however slightly to reach their incredible potential.

A. Soifer, *The Colorado Mathematical Olympiad and Further Explorations:*
From the Mountains of Colorado to the Peaks of Mathematics,
DOI 10.1007/978-0-387-75472-7_46, © Alexander Soifer, 2011

Reminiscences in Eight Parts

PART 1: *Gazette Telegraph*

"Math Olympiad contestants figure on five tough challenges" by Rick Ansorge, *Gazette Telegraph*, Thursday, April 22, 1993:

Math problems that make most people's eyes glaze over bring a sparkle to the eyes of Alexander Soifer.

Soifer – a math prodigy, Soviet émigré and professor at the University of Colorado at Colorado Springs – is the driving force behind the Colorado Mathematical Olympiad.

The 10th-annual event, which begins at 8:15 a.m. Friday in UCCS's Science Hall, will pit the minds of up to 1,100 junior and senior high school students against five maddeningly difficult problems devised by Soifer during spring break.

An A in calculus won't necessarily do them any good.

"There's perception that it's the A students, the ones who always say 'good morning' to their teachers and brush their teeth every night, who do the best," Soifer says.

"Not true."

One year, a C student took top honors.

First-year algebra students have as good a chance as anyone of carting home a scholarship, book, or high-tech calculator.

In 1985, ninth-grader David Hunter showed up his elders by winning the silver medal. He went on to become what Soifer calls the Olympiad's "all-time champ" by winning three gold medals in a row.

Hunter was Palmer High School's 1988 valedictorian, and graduated from Princeton University in 1992. Now 23, he

A. Soifer, *The Colorado Mathematical Olympiad and Further Explorations: From the Mountains of Colorado to the Peaks of Mathematics,* DOI 10.1007/978-0-387-75472-7_47, © Alexander Soifer, 2011

teaches math at a New Hampshire high school, but plans to study graduate-level math in a year or so.

Like tennis players, mathematicians do their best work when they're young.

"They peak early," Hunter said.

Hunter credits his parents – James and Louise – for giving him good genes and a healthy curiosity.

"They encouraged me to ask questions, and try to figure out the answers on my own," he said.

Louise Hunter recognized early that her son had smarts to grapple with abstract concepts. "He started reading at 3," she said. "He's always been very focused mentally."

Concentration, intuition and creativity are essential to just decipher Soifer's math problems, let alone solve them. No Olympiad participant to date has successfully solved all five.

Unlike some intelligence tests, however, the Olympiad is a fairly good predictor of success.

Russell Shaffer, the Olympiad's first winner, graduated from the Massachusetts Institute of Technology (M.I.T.) with a perfect 5.0 average, and went on to earn a doctorate in computer science from Princeton. He's now a researcher for MCI.

Soifer delights in telling such success stories because they closely mirror his own. As a youngster growing up in Moscow, he was aiming toward a music career when he began competing in and winning math contests. Soon, he was using muscles in his head that most people don't realize exist.

"It opened the word for me," he said.

PART 2: *Matt Fackelman*

You have already met Matt Fackelman in the history of year 6. He was the one who made me think of rewarding not only mathematical achievements of our Olympians, but also literary and perhaps even artistic ones. Why should we, the Colorado *Mathematical* Olympiad, reward student talents in literature and fine art? Simply because there are no literary or art Olympiads around. And also because every talent in young people ought to be treasured and supported, and if we can, we must do it!

Matt's essays (see a quotation in in the history of year 6) were deep yet humorous, and metaphysical above all. It was easy to decide to award Matt our Olympiad's first Literary Award. But what should I give him? After browsing through my private c. 8000 volume library, I chose for Matt the book *Metamagical Themas* by the Pulitzer Prize winner Douglas Hofstadter. My choice was quite good, as Matt Fackelman's May 22, 1989, letter shows:

Dear Mr. Soifer,

First of all I want to thank [you] for making such an enjoyable contest available to all high school students. Second I would like to thank you for the wonderful gift of *Metamagical Themas*. It is truly magical.

As I walked in to the Awards Ceremony, I sat down and immediately stood right back up. This is not a normal ritual with me, however, I saw that there were several copies of *Gödel, Escher, Bach* on the table and I got excited.

As you were passing out awards to your assistants, I saw that there were not only copies of *Gödel, Escher, Bach* but Hofstadter's second book as well. As I saw this book, I immediately became a zenith quantum of excitement, for I had been planning on getting the book for quite some time.

As you finished passing out these awards to your assistants, there was only one copy of *Metamagical Themas* left. I was somewhat disappointed because I didn't think that I had any chance of winning this last copy. However... the next thing I knew I was shaking your hand and you were presenting me with the book. At no other moment in my life have I been more completely satisfied.

I want to thank you and your assistants for making the Math Olympiad one of the greatest memories in my life. Thank you!! "You'll see me there," said the Cat, and vanished. Alice was not much surprised at this, she was getting so well used to queer things happening."

– Lewis Carroll

Your friend always
(in mathematics and metaphysics)

Matt Fackelman

Today Matt passes on the baton of creativity by teaching Film and TV courses at Palmer Ridge High School.

PART 3: *Brian Becker*

Brian E. Becker is a son of a late beloved colleague of mine, Psychology Professor Lee A. Becker, and Donna M. Becker, a lecturer in foreign languages. In 1990 Brian won second prize in the Colorado Mathematical Olympiad. On September 15, 1991, Brian shared with me the great news of being accepted by Stanford University:

Dear Professor Alexander Soifer:

Thank you very much for allowing myself and other young students the opportunity to enrich their lives at your Mathematical Olympiad. I believe that this contest is an excellent way for students to become interested in mathematics beyond the classroom. It has been an excellent opportunity to indulge and exercise the limits of my mind's logical and intuitive capacities. I only wish I would have started participating a few years earlier so that I might have been able to beat Matt Kahle![1] I am most disappointed that I will not be able to return next year. Your Olympiad is certainly a manifestation to all of fun and excitement that mathematics can create.

Now that you are surely beaming with delight about the success of your Olympiad, I feel it would be a good time to ask for

---

[1] Part 8 of this chapter is dedicated to Matt Kahle.

money! I would like my $300 check from the [second prize at the] 1990 Olympiad sent to Stanford University...

Thank you once again for holding such a wonderful contest at the students are always motivated to test and prove their intellect and win so many great prizes!

Sincerely,
Brian Becker

P.S. I plan to pursue my interests in mathematics at Stanford.

P.P.S. My calculators are wonderful – I've programmed my first one to play polkas!

Brian graduated from Stanford University with a master's degree in computer science. He worked for Oracle, Sun Microsystems. When his father, Lee, fell seriously ill, Brian moved back to Colorado Springs, where he still works for a variety of companies, including those in Silicon Valley. Brian holds a number of patents.

PART 4: *Charlie O'Keefe*

On May 2, 1998, two days after the Olympiad's Award Presentation, the silver medalist Charlie O'Keefe made it all worthwhile by his e-mail:

...I would like to thank you for putting on the Math Olympiad all these years. Math has always been one of my strengths, but seldom have I enjoyed working on problems as much as I have enjoyed the Olympiad's problems. They have a beautiful simplicity to them but are at the same time challenging.

Three years ago, when I first competed in the Colorado Olympiad, I approached it like an ordinary test. Admittedly, I am not a "show your work" kind of person (that has always been my biggest source of lost points – not showing enough work) and I certainly wasn't that day, but I still did well enough for a first honorable mention. I missed the award ceremony that year due to a track meet. The next year before the Olympiad, to my surprise, I came across the "First 10 Years" book in the school library. A look at that book was a great eye opener – I realized for the first time what kind of competition this really is – not a measure of speed or test taking ability or memorization of facts or anything else that the school system attempts to sharpen in students. This is a test of reasoning ability and creativity, which are much more valuable.

The next time around, I approached the test the way it ought to be approached – as an essay contest, and I wrote nice explanations for all the ones I solved (still couldn't get those last couple) I also packed more wisely for this one. While almost everyone else around me had a fancy graphing calculator along, I brought the HP scientific I won the year before, knowing that I might use it perhaps once or twice during the test for a simple multiplication here or there. In my other pocket, I had a tool that I have brought both years since then and never regretted: a box of colored pencils. I got first honorable mention again that year, but higher up on the list.

Interesting side note – I lived in Fort Collins many years ago, and I had a favorite babysitter that would teach me things about math and music. He went on to M.I.T. and his name is Adam Fedder. Imagine my surprise when, 10 years later while reading the 1st 10 years book, I saw his name on page 94 as a third prize winner!

... This year I almost got it right, except for a weak proof of 3b and a failed attempt at 5b. I came away very proud of my solution to 4a, though. Was my solution new, or did you have that one?

In my college search this year, after visiting several, I narrowed things down to Harvey Mudd in Southern California and the School of Mines. I was accepted by both, and then chose Mines. That scholarship will help!

Final thoughts: I have started reading my new copy of "Geometric Etudes" and I absolutely agree with you that these kinds of puzzles should be taught in school. If I practiced them that often, I would probably improve in many areas of thinking. If I were a teacher, I would base my whole curriculum around them! They are refreshingly fun to do, and unlike typical school math, produce a sense of accomplishment when you nail one!

If possible, could you send me the solutions for this year? I am especially anxious to see your solutions to 4a and *especially* 5b!

I wish you good luck in the coming years of the Olympiad, and maybe with some luck Combinatorics will catch on in public schools! I think that in the Colorado Math Olympiad I have found a new lifelong hobby.

Sincerely,

Charlie O'Keefe

PART 5: *David Hunter*

I met David in 1985 when as a ninth-grader he came to the Second Olympiad. There were no ties that year, just one first prize and one second prize. David won second prize and came in plaster on crutches to receive it. A year later, as a tenth-grader David won first prize, which in addition to gold medal included a scholarship to the International Summer Institute in gorgeous Southampton, Long Island, New York. There I got to really know this tall handsome young man with blond curly hair and blue eyes. We met daily for math, where in the international group of peers David did quite well, as well as on the volleyball court and the Atlantic Ocean beach. I knew David as a talented mathematician; the camp showed him also as a young man of highest integrity. Back in Colorado Springs I told his mother Louise,

– If you were to ever have any problems with Dave, I am standing in line to adopt him.
– Thank you! – was her reply.

In four years of competing, David won one silver and three gold medals, a record that still stands (although Bryce Herdt came close by winning three gold and one bronze medals; and the home-schooled Mark Heim by winning three gold medals out of three Olympiads he participated). Following his high school graduation, David was Princeton bound, when he surprised me yet again. This high school kid said:

– Before going to graduate school, I want to give back, to teach high school for two years.

And David kept this youthful promise! Upon a brilliant completion of Princeton University, for two years he taught at Alvirne

High School in Hudson, New Hampshire, before entering the Ph.D. statistics program at the University of Michigan. He is now a (tenured) Professor of Statistics at Pennsylvania State University.

Several times, when Dave came to Colorado Springs to visit his parents, he and I met for dinner and conversation. One year David returned to the Colorado Mathematical Olympiad as a speaker at the Award Presentation Ceremonies. It was unforgettable to see the three-time winner addressing the kids who came in his place!

David Hunter writes on January 1, 2009:

My first experience with the Colorado Math Olympiad and with meeting Dr. Soifer was in the Olympiad's second year in 1985 — can it really have been almost a quarter century ago? — and, like most of the participating students, I had been involved with math contests of various sorts throughout junior high school. But I still had no idea what to expect of the Olympiad. Four hours for five questions? It seemed like an awfully strange format for a contest to me at the time, but I grew to love the Olympiad and I still recall it with great fondness. Of course, four hours for five questions is a lot closer to what I do today as a statistics professor — thinking about a single problem for days or weeks at a time, then carefully writing up what I've done — than the quick, tidy quizzes that most people think of when they think "math"; but that's not what I like about the Olympiad.

The point is that the Olympiad is a celebration of real mathematics, of elegance and creativity in reasoning. This is not a common occurrence in high school! I believe that there is something wonderful about an event at which hundreds of students voluntarily spend a whole day on five math problems (and then return a week later to hear the solutions and see who did well!).

I now have a job in which I do math a lot of the time, so the fact that I've always loved math means that I have a bit of built-in job satisfaction. As one of my jobs in the statistics department at Penn State, I am the chair of our undergraduate program, which means that I occasionally get to advice students prior to and during their undergraduate academic careers. My main piece of advice is this: Take as much math as you can. I was a math major in college and have always believed that a degree in math is an excellent basis for almost any career: It proves that a student can think. It lacks the vocational flavor of a degree like engineering or computer science, in which students learn particular skills; but such things can often be acquired later, either in graduate school or on the job, where it is often the case that job-specific knowledge is ultra-specialized anyway.

But I digress. When he contacted me about writing this essay, Dr. Soifer reminded me of a quotation of mine that appeared in the *Gazette Telegraph* one of the years I participated in the Olympiad, something along the lines of "I am not a nerd." This made me smile, because recently I have been known to announce to a lecture hall of hundreds of students in an introductory statistics class, when the situation calls for it, that I am a math nerd. Partly I do this to get their attention and elicit a bit of a laugh, and the students can tell exactly what I mean: I love mathematics, and even though few of them will ever be able to say that — my intro stats class is not generally a Math Olympiad kind of crowd — they can appreciate at least that I am trying to share my enthusiasm for statistics with them. Just like Dr. Soifer has been sharing his enthusiasm for mathematics through the Olympiad for all these years.

PART 6: *Gideon Yaffe*

Gideon Yaffe won third prizes as a sophomore in 1986 and as a junior in 1987, and first prize as a senior in 1988 (Gideon and David Hunter were two co-winners that year). Gideon also won the 1988 Creativity Award for coming up with a new idea in treating 6 points in a triangle. Talented in arts and sciences, Gideon has had an exciting life full of creative achievements.

Gideon writes on November 11, 2008:

Dear Alexander,

Nice to hear from you. I've had news of you now and again from Paul Zeitz. I'm glad to hear that you are thriving.

I wandered from major to major at Harvard. I never actually majored in drama, although I had intended to when I went to college. I acted in some plays, but never majored. I majored in math for a time, in biology for a time, in applied math for a time, and in various blank-and-blank combinations. In the end, I majored in philosophy largely because I wanted to write a senior thesis in the area. My initial interest in philosophy came from my interest in math. I was very taken by logic and by the way in which formal argument could be used to make progress on philosophical questions. After some time living in the Bay Area and working some stupid jobs, I went to graduate school in philosophy at Stanford and finished my Ph.D. in 1998.

I wrote my dissertation on the free will problem and on the seventeenth century philosopher John Locke's answer to it. In 1999, I started on the tenure track at the University of Southern California [USC].

I published two books between 2000 and 2004, one on Locke and one on a fairly obscure 18th Century philosopher named Thomas Reid. Also in that period I developed interests in legal issues, particularly the application of philosophical thought about free will and responsibility to criminal law. In the 2004-05 school year, I was a law student at USC, and two years later the law school hired me part time. So now I split my time between the law school and the philosophy department. For the last few years, I've been publishing in philosophy of law and I'm currently working on a book about criminal attempts – the crime of attempting a crime. I'm also involved with the Macarthur Foundation's law and neuroscience project. The Macarthur has dedicated $10 million to study the use of neuroscience to the law and I'm one of the people developing projects for that.

My family doesn't live in Colorado Springs anymore, so it's not on my usual trek any longer, unfortunately. However, if I find myself there, I'll be sure to look you up.

Thanks for the Olympiad. It made a big difference to me when I was in high school. You might remember that the year that David Hunter and I won it, I also won a creativity prize. That probably meant more to me than the victory in the Olympiad itself. It made me realize that the brain that I used to do math could be used for lots of things, anything that required creativity, and so it prompted me to try out lots of things in college.

Best,

– Gideon

PART 7: *Aaron Parsons*

The best teachers are usually found in major metropolitan centers, where schools are well funded and cared for by an enlightened populous. The small town of Rangely in the very Northwestern corner of the state of Colorado near the Colorado-Utah border did not hold promise of mathematical inspiration for their small student body – unless a miracle were to happen. It did! The miracle's name was Melvin Oliver, mathematics teacher at Rangely High School.

Mel must have practiced magic, for nearly all the students he brought to the Colorado Mathematical Olympiad through many years, won various awards. (We typically presented awards to no more than 20% of contestants.) In most years, Melvin Oliver and his students traveled twice to Colorado Springs (a 13-hour round trip by car), to attend both the Olympiad and a week later the Award Presentation. Among the many of Mel's fine Olympians, one stood out: Aaron Parsons, who won the silver medal as a junior in 1997, and the gold medal in 1998. All this Aaron achieved while successfully competing in track and field – sprint to be precise – at the state's highest level, placing third in Colorado. And Aaron was rewarded by an improbable journey, from the town of Rangely to Harvard University as a mathematics major!

When invited to Harvard, Aaron shared the good news with me in his July 25, 1998, letter:

Dear Professor Soifer,

I would first like to say that I have enjoyed your contest throughout my high school years. It is one of the few contests in Colorado that places all students on an equal level and allows each to test his/her merit through problem-solving skills and critical thinking. This contest, to me, comes closer to capturing the essence of mathematics than any other contest I have ever taken. Thank you very much for putting together such a remarkable contest for so many years.

I would also like to thank you for helping provide for my college education by offering scholarships to the winners of the contest. I have been accepted to Harvard University, and will pursue a degree in mathematics and physics there this coming year...

Thank you again for your wonderful math contest. I am sure you will be seeing a lot more of my favorite teacher, Mel Oliver, and the Rangely Math Club.

Success had no effect on this modest talented cheerful young man. Ten years passed. I contacted Aaron in December 2008 and asked him to share his experiences during and after the Olympiad. On December 11, 2008 Aaron replied:

Dear Prof. Soifer,

How wonderful to hear from you! Sorry that I've been slow to return the message you left on my guestbook (I only check it infrequently). I would be more than happy to provide a short note about myself and CMO:

I participated in the Colorado Math Olympiad [CMO] from 1994 to 1998, representing Rangely High School [RHS] and coached by the generous and committed Melvin Oliver, who single-handedly developed and supported the math program at RHS. At a time when most math contests focused on speed, numbers, and arithmetic tricks, CMO stood out as something completely different.

The first time I took a CMO test, I was flabbergasted – I was so tuned to the "other" type of contest, I felt I could hardly solve a single problem! No one on our team qualified for the awards ceremony that year, and one of the problems just drove me nuts – one about polygons of unit area on a grid. Resolving to qualify for the "answer session" next year, I set to work. I did manage an "Honorable Mention" the next year, and after visiting UCCS and meeting Prof. Soifer for the annual recapitulation of the contest, both Mel and I came away with a new understanding of a broader, more abstract, and altogether much more fun side of mathematics than we had previously seen.

I distinctly remember the following year, when, having qualified for the awards ceremony, I found myself unfortunately needing to skip the ceremony in order to participate in a state track meet (held just an hour away). I met with Prof. Soifer to excuse myself and to apologize for necessity of my departure. "Oh, that's quite alright," he said. "You know, I was a sprinter myself in high school." And a good one at that, as I found out. I began to understand his bounding energy in front of his students, to guess at a joie d' vivre that could be expressed both academically and athletically.

After graduating from high school, I studied physics and mathematics (and ran track) at Harvard. There, I discovered that math – the real math that mathematicians do – was really much more like the bounding, gleeful CMO math than any other math I had been exposed to. I grew much better versed in mathematical reasoning, but it still wasn't until my second year away at college that the solution to that demonic polygon problem finally came to me. Liberated at long last, I moved on to astrophysics. I am currently finishing my doctorate at [the University of California] Berkeley, working to discover the first stars that formed in the universe 10 billion years ago.

– Aaron Parsons

PART 8: *Matthew Kahle* [2]

I first met Matt, or as he used to sign his name, M@, on April 17, 1987. He was an eighth-grader, and his mom brought him to the Fourth Colorado Mathematical Olympiad. Matt was holding in his hands his baby sister. A month later I met the three in Washington DC, where Matt was on the Colorado team for the National Competition of MATHCOOUNTS (and I was a National judge[3]). The following year, in 1988, Matt won third prize in the 1988 Colorado Mathematical Olympiad. His grades in school were low; even his geometry grade was a C. I asked his fine math teacher Judy Williamson,

– Why did Matt get a C in your geometry class?
– Because he does homework only when he likes problems. I have 40 students, and cannot possibly provide individual instruction.

I had no answer for Judy, but did have one for Matt. I told him:

– Any time you need a math problem, or simply to talk, come to me.

---

[2] Part of my reflections about Matthew Kahle has first appeared in the new Springer edition of *How Does One Cut a Triangle?* In 2009 [S8].

[3] In fact, I had been a national judge of MATHCOUNTS for 12 years, 1984–1995.

And Matt came: he took my university problem solving class as a high school freshman. He was destined to do better in the Olympiad than his 1988 third prize: Matthew won first prize and the gold medal in both 1990 and 1991 Colorado Mathematical Olympiads, and proved that he was the best high school mathematician in Colorado at the time. However, this did not open for him the doors of my University of Colorado. The reason? A low 1.9 grade point average of high school grades (about C–).

What was I to do? I wrote a letter to the Director of Admissions and Records of my university, Randy Kouba. I argued:

– We, the University of Colorado, need Matt more than he needs us. One day we would be proud to have such a brilliant alumni.

All this was to no avail. Randy Kouba replied:

– We have rules. Let him go to Pikes Peak Community [2-year] College and improve his grade point average. Then he could try us again.

"Then" has not happened, because Colorado State University at Fort Collins offered Matt a full scholarship and Matt got his bachelor's and master degrees there. Matt then earned his Ph.D. degree in mathematics from the University of Washington, Seattle. He was awarded a postdoctoral fellowship at Stanford University, and in the fall of 2010 is starting a postdoctoral fellowship at the historic Institute for Advanced Study in Princeton, at one point the seat of Albert Einstein.

My university – more precisely Chancellor Pam Shockley and Arts and Sciences Dean Tom Christensen – eventually approved an "admissions window," i.e., exceptions in admissions for students who won any medal in the Colorado Mathematical Olympiad during their high school careers. In this, the university matched for the Math Olympiad the exceptions it had long granted to American Olympic sportsmen (Headquarters of the USA Olympic Committee and the training center are located in Colorado Springs). My thesis that we ought to reward brains at least as much as we reward arms and legs has finally celebrated its victory!

Perhaps more than any contestant, Matt absorbed the Olympiad spirit. He competed in five Olympiads, and he came back seven times to serve as the Olympiad's judge, to participate in passing the flaming Olympic torch to the next generations of young mathematicians.

In order to help with judging, Matt traveled 120 miles each way from Fort Collins – you see him here in the 2001 photograph during one such judging visit. In 2008, Matt traveled 1200 miles each way to come from Stanford University to Colorado Springs.

Matthew Kahle (right) and Alexander Soifer on April 21, 2001 during the eighteenth Colorado Mathematical Olympiad.

In his December 2008 essay [Ka] that appears in *Geombinatorics* in January 2009 and is dedicated to my round numbered birthday, Matt speaks of what he learned from me before improving one of the theorems of my book *How Does One Cut a Triangle?*:

> Perhaps the most important thing that I learned from him is that we are free to ask our own mathematical questions and pursue our own interests, that we can trust our sense of aesthetics and our intuition. This was a very empowering idea to me, particularly when I was young and struggling with boredom and frustration with school.

I see no better way to end this volume than with Matt's essay written on December 30, 2008, especially for this book:

> Professor Alexander "Sasha" Soifer told me that for the new edition of the Colorado Mathematical Olympiad book, he wanted to include a chapter written by past winners, about the role of the Olympiad in their lives, their view of math, and their future careers. I am very happy to contribute an essay for this.

I competed in many math competitions growing up – in several statewide competitions sponsored by the local universities, as well as in well known national contests – MathCounts, USAMO [USA Mathematics Olympiad], Putnam, etc. The first thing that comes to mind was that these competitions helped me realize that I was good at math, and perhaps even more importantly, they helped realize how much I enjoy it. And they helped instill me with a sense of confidence when I was an angst filled adolescent with low self esteem. So for those things I will always be very grateful.

Given the topic of this essay, however, I want to emphasize that there are many features of the CMO [Colorado Mathematical Olympiad] that distinguish it in my mind from the other contests which I competed in. First, the format itself was different from most competitions – five questions and four hours! It is true that there are other Olympiad style contests that I would come across years later (USAMO, Putnam, ...), but I first sat for the CMO as an 8th grader, when our teacher and coach Betty Daniels took a few of us from the middle school to UCCS [University of Colorado at Colorado Springs] for the Olympiad.

I had never seen such an exam before! I remember Sasha coming around to the rooms at the beginning of the day, asking if anyone had questions. I think I asked him if we could use calculators. He smiled impishly, shrugged his shoulders and said, "Sure, why not?" (Perhaps, needless to say, calculators weren't much help.) I am pretty sure that I spent the whole four hours every year I took it, even in eighth grade, and I never solved all of the problems. But it was not an ordeal for me to take a four hour math test; I could leave any time I wanted. I stayed until the end because I was having fun. This might be one of the most important qualities that separate the CMO in my mind – the spirit of fun, and even the sense of humor.

Not to say that the Olympiad was not serious business, not at all. In fact I would say it was some of the hardest tests that I ever took, and I am more proud of my performances in the CMO than in any other math contest. But it was always a pleasure to work on the problems. And I would continue thinking about the ones that I hadn't gotten all week, and then come back for the award ceremony to see the answers revealed. In hindsight,

Soifer's presentations at the award ceremony feel like some of the first real math lectures I ever saw. They were so far beyond what they were teaching us in school, so different in style, yet somehow accessible at the same time. It felt like he was giving us a peek of an entire other world that we had never been exposed to.

I grew up to be a mathematician, but my road was a little bit of a long and winding one. In short, I was always much more excited about mathematics itself than I ever was about school, so much so that I barely made it out of high school and dropped out of college twice. But after a few false starts I set my sights on getting a math Ph.D., which I finished at the University of Washington in 2007. Since then, I've been a postdoctoral fellow at Stanford University. I greatly enjoy both research and teaching, and I hope to find a tenure track job as a professor that will allow me to continue both.

I think the Olympiad influenced me in many ways. It introduced me to the idea that there are many more math problems unsolved than there are solved. (I remember Professor Soifer saying, "To nearest percent, 0% of all math problems are solved."). It helped me find out that I am capable of obsessing about math problems for hours, or days, or longer. (Needless to say, this is an important trait for a math researcher.) And that which problems we work on, and how we work on them, is not only a matter of ability, but of aesthetics and taste. (The CMO awarded special prizes for creative solutions, as well as literary prizes for clever poems and stories.) I think it also helped me realize how much I enjoy just talking with people about math – some of the first "serious" math conversations I remember having with Soifer.

I appreciate that Sasha treated me as a friend and a peer even when he first met me as a 14 year old, and I have stayed in touch with him and connected to the Olympiad since then. I have contributed three research articles so far to his journal *Geombinatorics*, and helped judge the CMO several times. I was honored to be able to help judge for the 25th annual CMO last year. I think Sasha and I are just kindred spirits. We may naturally have some similar tastes mathematically, but I also think we recognized and appreciated the mischief and humor in each other's eyes from when we first met. For some reason, right now I am remembering winning the Olympiad, and shaking hands with Professor Soifer and one of the deans, and Soifer turning to the dean and saying, "See! This is why we should admit C-students!"

This is, by necessity, a bit of a personal note, and the particulars of my story might be unique. But something that I really appreciate, especially after having seen things from the other side as a judge, is just how many students have been affected. Several hundred students compete annually, and prizes are awarded to over a hundred students almost every year. Now multiply that by 25 years. So on behalf of all the lives and minds you've touched, many thanks Sasha. And here's to 25 more years of the Colorado Mathematical Olympiad!

Farewell to the Reader

The Colorado Mathematical Olympiad is on its 28th year. It has become a lifelong commitment for me. Even when I lived and worked at Princeton University (2002-2004 and 2006-2007), I flew back to Colorado Springs to run the Olympiad and a week later the Award Presentation. The Olympiad has been my way to pass the mathematical baton to kids of Colorado and to the next generations of creative thinkers in general and mathematicians in particular. It has also been my way to show my colleagues that even one possessed, inspired (crazy? :-) person can pull off such a major event. I hope the Olympiad and this book will sow seeds of many other Olympiads. (According to Paul Zeitz, it has already inspired the Bay Area Mathematical Olympiad.)

Thank you for holding my book in your hands. I welcome your ideas, comments, solutions of problems presented here and new problems you may create. I hope to offer you soon a continuation of this book entitled something like *Colorado Mathematical Olympiad: The Third Decade and Further Explorations*.

As Paul Erdős used to say at the end of his lectures, "everything comes to an end," and so has this book. However, if you are inclined to continue your explorations of mathematics with me, I have good news for you. This book is one of my eight books that Springer has or is soon going to publish.

If you are receptive to a visual appeal of geometry, you may wish to read the expanded edition of *How Does One Cut a Triangle?* [S13]. Its first edition [S2] was published in 1990. Moreover, this book offers a glimpse of "real" mathematics, a demonstration of synthesis, where ideas from various branches of mathematics work together to achieve a geometric result.

A. Soifer, *The Colorado Mathematical Olympiad and Further Explorations: From the Mountains of Colorado to the Peaks of Mathematics*, DOI 10.1007/978-0-387-75472-7, © Alexander Soifer 2011

The expanded edition of *Geometric Etudes in Combinatorial Mathematics* [S14] has the dual goal of showing how geometric insight does wonders in service to combinatorics. Its first edition [BS] came out in 1991.

Election day, November 4, 2008 (the "yes-we-can" day), saw the release of the book I dreamed of and worked on for 18 years, *The Mathematical Coloring Book: Mathematics of Coloring and the Colorful Life of Its Creators* [S11]. This voluminous book offers the beautiful mathematics of coloring (so-called *Ramsey Theory*), and historical investigations into the lives of mathematicians, from the Nazi era in Germany to the Netherlands, devastated by World War II. The history allowed me to pose questions that have not lost their urgency today, such as the role of a scholar in tyranny. The book presents the aesthetics of mathematics as an art, philosophy of its foundations, and psychology of mathematical and historical discovery. It also shows that mathematics can become a genre of literature. The Nobel laureate Boris Pasternak expressed my goals in this book better and more concisely than I could – great poets do magic with words:

I bring here all: what have I lived thru,
And that what keeps my soul alive,
My rectitude and aspirations,
And what have seen my own eyes.[1]

My next book will not include mathematics. However, the great twentieth century mathematician will be the hero of the book, which will be entitled *Life and Fate: In Search of Van der Waerden* [S15]. I hope it will be published in 2011.

A book of open problems of the legendary mathematician Paul Erdős will come next, likely in 2012: *Problems of pgom Erdős* [ES2]. I would not have attempted to write it, but in 1990 Paul asked me to help him in the endeavor, and we had been working on this book together for years, from 1990 to his passing in 1996. This will be our joint book... As you may know, Paul Erdős (1913-1996) was the greatest problem creator of all time. You will be able to work on his problems because no knowledge is required to understand many of them. Better yet, many of Erdős's problems allow young mathematicians to advance, by finding at least partial solutions.

---

[1] [Pas], Translated especially for *The Mathematical Coloring Book* [S11] by Ilya Hoffman.

The book after Erdős's problems would be either *The Art on the Frontier of Cultures: The Fang People of West Equatorial Africa and Their Neighbors* or *Memory in Flashback: On Both Sides of the Atlantic*. The former will be a result of my ongoing study of African Art and culture, inspired by the great anthropologist and my friend James W. Fernandez, who lived with the Fang during colonial times 1958-1960. The latter will be a collection of humorous and noteworthy vignettes sketching moments of my life.

Having finished this book, you have become my alumnus, a title that carries responsibility to stay in touch, to send me your most enjoyable solutions, and your new problems, suggestions, and ideas. Be rest assured: I will always be delighted to hear from you!

References

[AM] Abbot, H. L., and Moser, L., Sum-free sets of integers, *Acta Arith.*, 11 (1966), 393-396.

[BS] V. Boltyanski and A. Soifer. *Geometric Etudes in Combinatorial Mathematics*, Center for Excellence in Mathematical Education, Colorado Springs, 1991.

[CFG] H. T. Croft, K. J. Falconer, and R. K. Guy, *Unsolved Problems in Geometry*, Springer, New York, 1991.

[DGK] Danzer, L., Grünbaum, B., and Klee, V., Helly's Theorem and Its Relatives, in *Convexity, Proceedings of Symposia in Pure Mathematics,* vol. VII, Amer. Math. Soc., Providence, 1963, 101-180.

[E0] Erdős, P., On Sets of Distances of n Points, *Amer. Math. Monthly* 53(5), 1946, 248-250.

[E1] Erdős, P., Combinatorial problems in geometry and number theory, *Relations between combinatorics and other parts of mathematics, Proc. Sympos. Pure Math.*, Ohio State Univ., Columbus Ohio (1978); *Proc. Sympos. Pure Math.*, XXXIV, Amer. Math. Soc., Providence, RI (1979), 149-162.

[E2] Erdős, P., Problems and results in combinatorial geometry, *Discrete Geometry and Convexity, Annals of the New York Academy of Sciences*, 440, The New York Academy of Sciences, New York (1985), 1-11.

[E3] Erdős, P., Problems and results on finite and infinite combinatorial analysis, *Infinite and Finite Sets (Colloq., Keszthely, 1973), vol. I: Colloq. Math. Soc. János Bolyai*, Vol. 10, 403-424, North-Holland, Amsterdam, 1975.

[E4] Erdős, P., Some recent problems and results in graph theory, combinatorics and number theory, *Proceedings of the Seventh Southeastern Conference on Combinatorics, Graph Theory, and Computing (Louisiana State Univ., Baton Rouge, 1976), Congress. Numer.* XVII, 3-14, Utilitas Math., Winnipeg, 1976.

[E5] Erdős, P., Some recent problems and results in graph theory, *Discrete Math.* 164 (1997), 81-85.

[E6]   Erdős, P., Combinatorial problems in geometry, *Math. Chronicle* 12 (1983), 35-54.

[EGMRSS] P. Erdős, R. L. Graham, P. Montgomery, B. L. Rothschild, J. H. Spencer, and E. G. Straus, Euclidean Ramsey Theorems I, II, and III, *J Combin. Theory Ser. A*, 14 (1973) 341-363;*Coll. Math. Soc. Janos Bolyai*, 10 (1973) North Holland, Amsterdam (1975), 529-558 and 559-584.

[ES1]   Erdős, P., and Soifer, A., Squares in a Square, *Geombinatorics* IV(4), 1995, 110-114.

[ES2]   Erdős, P., and Soifer, A., *Problems of pgom Erdős*, Springer, New York, 2012, to appear.

[ESz]   Erdős, P., and Szekeres, G., A combinatorial problem in geometry, *Compositio Math* **2** (1935). 463–470.

[Eu]   Euler, L., Solutio problematic ad geometriam situs pertinentis, *Comment. Academiae Sci. Imp. Petropolitanae* 8 (1736), 128-140.

[F]   Freud, S., *Civilization and Its Discontents*, Dover, New York, 1994.

[GT]   G.A. Galperin and A.K. Tolpygo. *Moscow Mathematical Olympiads*, edited and with introduction by A.N. Kolmogorov, Prosvetshenie, Moscow, 1986 (Russian).

[G1]   M. Gardner, The celebrated four-color map problem of topology, *Scientific American*, 205 (Sep. 1960), 218-226.

[G2]   M. Gardner, A new collection of brain teasers, *Scientific American*, 206 (Oct. 1960), 172-180.

[Gr1]   R. L. Graham, *Rudiments of Ramsey Theory*, American Mathematical Society, Providence, 1981.

[Gr2]   R. L. Graham, Some of my favorite problems of Ramsey Theory. In B. Landman et al. (eds.), *Proceedings of the 'Integer Conference 2005' in Celebration of the 70th Birthday of Ronald Graham. Georgia, USA, October 27-30, 2005*, Walter de Gruyter, Berlin, 2007, 229-236.

[GG]   R. E. Greenwood and A. M. Gleason, Combinatorial relations and chromatic graphs, *Canad. J. Math.*, 7 (1955), 1-7.

[GRS]   R. L. Graham, B. L. Rothschild, and J. H. Spencer. *Ramsey Theory*, second edition, John Wiley & Sons, New York, 1990.

[Grü]   Grünbaum, B., A proof of Vázsonyi's conjecture, *Bull. Res. Council Israel, Sect. A* 6 (1956), 77-78.

[H1]   H. Hadwiger, Uberdeckung des euklidischen Raum durch kongruente Mengen, *Portugaliae Math.*, 4 (1945), 238-242.

[H2]   H. Hadwiger, Ungelöste Probleme, Nr. 11, *Elemente der Mathematik*, 16 (1961) 103-104.

[Har]   Harary, F., Is Snaky a Winner? *Geombinatorics* II(4), 1993, 79-82.

[Ha]   Harborth, H., Lösung zu Problem 664A, *Elemente Math.* 29 (1974), 14-15.

[He]   Heppes, A., Beweis einer Vermutung von A. Vársonyi, *Acta Math. Acad. Sci. Hungar.* 7 (1956), 463-466.

[Ho] Hofmannsthal, H. von, *Buch der Freunde*, Leipzig, Insel, 1922. Translation into English by Douglas Robertson: http://shirtysleeves.blogspot.com/2008/04/translation-of-buch-der-freunde-by-hugo.html

[J1] Januszewski, J., Packing similar rectangles in a rectangle, *Geombinatorics* XII(1), 1992, 24-30.

[J2] Januszewski, J., Optimal translative packing of homothetic triangles, *Studia Scientiarum Mathematicarum Hungarica* (in print), DOI: 10.1556/SScMath.2009.1083

[JKST] Jelínek, V., Kyncl, J., Stolar, R., and Valla, T., V., Monochromatic triangles in two-colored plane, *Combinatorica* (submitted March 2006); arXiv:math.CO/0701940v1, Jan. 31, 2007.

[Ka] Kahle, M., Points in a triangle forcing small triangles, *Geombinatorics* XVIII(3), 2009, 114-128.

[KK] J. Kahn and G. Kalai, A counterexample to Borsuk's conjecture, *Bull. Amer. Math. Soc.* 29 (1993), 60-62.

[K] Kempe, A. B., On the Geographical Problem of the Four Colours, *American Journal of Mathematics* II (1879), 193-200.

[KW] V. Klee and S. Wagon, *Old and New Unsolved Problems in Plane Geometry and Number Theory*, Mathematical Association of America, 1991.

[Ku] Kupitz, Y. S., On the maximal number of appearances of the minimal distance among n points in the plane, in Böröczky et al. (eds), *Intuitive Geometry* (Szeged, 1991), *Colloq. Math. Soc. János Bolyai* 63 (1994), 217-244.

[MoM] Moon, J. W., and Moser L., Some packing and covering problems, *Colloq. Math.* XVII, 1967, 103-110.

[MM] L. Moser and W. Moser, Solution to Problem 10, *Can. Math. Bull.*, 4, (1961), 197-189.

[N] Novotný, P., A note on packing clones, *Geombinatorics* XI(1), 29-30.

[O'H] P. J. O'Halloran, editor. *World Compendium of Mathematics Competitions*, Australian Mathematics Foundation, Ltd. 1992.

[O] Ore, Ø, Note on Hamilton circuits, *Amer. Math. Monthly* 67 (1960), 55.

[Pas] Pasternak, B., Volny (Waves) in *The Second Birth*, Sovetsky Pisatel', Moscow, 1934 (Russian).

[R] Rado, R., Note on combinatorial analysis, *Proc. London Math. Soc.*, 48 (1945), 122-160.

[Ri] Richardson, T.J., Optimal packing of similar triangles, J. Combin. Theory, Ser. A 69 (1995), 288-300.

[SCH] I. Schur, Über Die Kongruenz: $x^m + y^m \equiv z^m$ (mod. p). *Jber. Deutsch. Math. – Verein*, 25 (1916), 114-117.

[Sha] L. E. Shader, All right triangles are Ramsey in E^2!, J. *Combin. Theory*, (A) 20 (1976), 385 - 389.

[S1] Soifer, A., *Mathematics as Problem Solving*, Center for Excellence in Mathematical Education, Colorado Springs, 1987.

[S2] Soifer, A., *How Does One Cut a Triangle?* Center for Excellence in Mathematical Education, Colorado Springs, 1990.

[S3] Soifer, A., Chromatic number of the plane: A historical essay, *Geombinatorics*, I(3) (1991), 13-15.

[S4] Soifer, A., Triangles in a three-colored plane, *Geombinatorics*, I(2) (1991), 13-14 and I(4) (1992), 21.

[S5] Soifer, A., Creating a new generation of problems for mathematical Olympiads, *Mathematics Competitions* 6(1), 1993, 60-74.

[S6] Soifer, A., *Life and Fate: In Search of Van der Waerden,* Springer, New York, 2011, to appear.

[S7] Soifer, A., Squares in a square II, *Geombinatorics* V(3), 1995, 121.

[S8] Soifer, A., Discs in a disc, *Geombinatorics* VII(4), 1998, 139-140.

[S9] Soifer, A., Packing clones in convex figures, *Geombinatorics* VIII(1), 1998, 166-158.

[S10] Soifer, A., From Squares in a Square to Triangles in a Triangle: A Ride on a Mathematical Train of Thought, *Mathematics Competitions* 13(2), 2000, 16-33.

[S11] Soifer, A., *The Mathematical Coloring Book: Mathematics of Coloring and the Colorful Life of Its Creators*, Springer, New York, 2009.

[S12] Soifer, A., *Mathematics as Problem Solving*, Springer, 2nd expanded ed., New York, 2009.

[S13] Soifer, A., *How Does One Cut a Triangle?* 2nd expanded ed., Springer, New York, 2009.

[S14] Soifer, A., *Geometric Etudes in Combinatorial Mathematics*, 2nd expanded ed., Springer, New York, 2010.

[S15] Soifer, A., *Life and Fate: In Search of Van der Waerden*, Springer, New York, to appear in 2011.

[S16] Soifer, A., Dvenadzat' Let "Geombinatoriki" (Dozen Years of "Geombinatorics"), *Matematicheskoye Prosveshchenie* (*Mathematical Enlightenment*), Ser. 3, issue 8 (2004), 132-135 (Russian).

[SL] Soifer, A., and Lozansky, E., Pigeons in every pigeonhole, *Quantum*, Jan. 1990, 25-28, 32.

[SS] Soifer, A., and Slobodnik, S. G., Problem M236, *Kvant* 12 (1973), p. 29 (Russian).

[St] Straczewicz, S., Su un probléme géométrique de P. Erdős, *Bull. Acad. Pol. Sci., Cl. III* 5 (1957), 39-40.

[TT] S. L. Tabachnikov and A. L. Toom. *Pouchitel'nye Igry (Didactic Games)*. Academy of Pedagogical Sciences of the USSR, Moscow, 1987 (Russian).

[T] Tait, P. G., Note on a theorem in the geometry of position, *Trans. Roy. Soc. Edinb.* 29 (1880), 657-660 & plate XVI. Reprinted: *Scientific Papers* 1(54), 1898-1900, 408-411 & plate X.

[TW] Taylor, R., and Wiles, A., Ring-theoretic properties of certain Hecke algebras, *Annals Math.* 142 (1995), 553-572.

[W] Wiles, A., Modular elliptic curves and Fermat's Last Theorem, *Annals Math.* 142 (1995), 443-551.

Index of Names

Index of Terms